Healthcare Services in the Metaverse

This book focuses on game theory approaches utilized on various domains viz., IoT, blockchain and their applications to biomedical and healthcare services. The book bridges the gap between radiologists and Artificial Intelligence (AI)-driven automated systems by investigating various techniques such as game theoretic approach, blockchain technology basically utilized for security, and IoT applied on metaverse.

Healthcare Services in the Metaverse: Game Theory, AI, IoT, and Blockchain, identifies the potential areas where game theory and block chain techniques can be harnessed in the metaverse. The book discusses the integration of virtual reality (VR) with augmented reality to identify the new emerging techniques in healthcare to metaverse, where doctors and/or patients can see any kind of operation in the VR metaverse. The authors use game theoretical and blockchain approaches to understand healthcare issues, with the aim of utilizing different technologies of metaverse platform for health informatics.

This book is written to help healthcare practitioners and individuals across academia and research, as well as for those who work in biomedical, Internet of Things (IoT), Artificial Intelligence (AI), metaverse, VR, blockchain, and related technologies.

Emerging Trends in Biomedical Technologies and Health informatics

Subhendu Kumar Pani, Orissa Engineering College, Bhubaneswar, Orissa, India
Sujata Dash, North Orissa University, Baripada, India
Sunil Vadera, University of Salford, Salford, UK

Everyday Technologies in Healthcare
Chhabi Rani Panigrahi, Bibudhendu Pati, Mamata Rath, Rajkumar Buyya

Biomedical Signal Processing for Healthcare Applications
Varun Bajaj, G R Sinha, Chinmay Chakraborty

Deep Learning in Biomedical and Health Informatics
M. Jabbar, Ajith Abraham, Onur Dogan, Ana Madureira, Sanju Tiwar

Computational Approaches in Biotechnology and Bioinformatics
Pranav Deepak Pathak, Roshani Raut, Sebastian Jaramillo-Isaza, Padnya Borkar, and Rutvij H. Jhaveri

Computational Approaches in Biomaterials and Biomedical Engineering Applications
Pranav Deepak Pathak, Roshani Raut, Sebastian Jaramillo-Isaza, Padnya Borkar, and Rutvij H. Jhaveri

Healthcare Services in the Metaverse: Game Theory, AI, IoT, and Blockchain
Saurav Mallik, Anjan Bandyopadhyay, Ruifeng Hu, Pawan Kumar Singh, Soumadip Ghosh, and Sujata Swain

For more information about this series, please visit: https://www.routledge.com/Emerging-Trends-in-Biomedical-Technologies-and-Health-informatics-series/book-series/ETBTHI

Healthcare Services in the Metaverse

Game Theory, AI, IoT, and Blockchain

Edited by
Saurav Mallik,
Anjan Bandyopadhyay,
Ruifeng Hu,
Pawan Kumar Singh,
Soumadip Ghosh, and
Sujata Swain

CRC Press is an imprint of the
Taylor & Francis Group, an informa business

Designed cover image: Shutterstock

First edition published 2024
by CRC Press, 2385 NW Executive Center Drive, Suite 320, Boca Raton FL 33431

and by CRC Press
4 Park Square, Milton Park, Abingdon, Oxon, OX14 4RN

CRC Press is an imprint of Taylor & Francis Group, LLC

© 2024 Taylor & Francis Group, LLC

Reasonable efforts have been made to publish reliable data and information, but the author and publisher cannot assume responsibility for the validity of all materials or the consequences of their use. The authors and publishers have attempted to trace the copyright holders of all material reproduced in this publication and apologize to copyright holders if permission to publish in this form has not been obtained. If any copyright material has not been acknowledged please write and let us know so we may rectify in any future reprint.

Except as permitted under U.S. Copyright Law, no part of this book may be reprinted, reproduced, transmitted, or utilized in any form by any electronic, mechanical, or other means, now known or hereafter invented, including photocopying, microfilming, and recording, or in any information storage or retrieval system, without written permission from the publishers.

For permission to photocopy or use material electronically from this work, access www.copyright.com or contact the Copyright Clearance Center, Inc. (CCC), 222 Rosewood Drive, Danvers, MA 01923, 978-750-8400. For works that are not available on CCC please contact mpkbookspermissions@tandf.co.uk

Trademark notice: Product or corporate names may be trademarks or registered trademarks and are used only for identification and explanation without intent to infringe.

ISBN: 978-1-032-58080-7 (hbk)
ISBN: 978-1-032-58246-7 (pbk)
ISBN: 978-1-003-44925-6 (ebk)

DOI: 10.1201/9781003449256

Typeset in Times
by KnowledgeWorks Global Ltd.

Contents

About the Editors

Dr. Saurav Mallik is currently working as a research scientist in the Department of Pharmacology and Toxicology at University of Arizona, USA. Previously, he worked as a postdoctoral fellow at Harvard University, MA, USA. Previously, he worked as a postdoctoral fellow in the Center of Precision Health, Department of School of Biomedical Informatics, University of Texas Health Science Center at Houston, TX, USA, and in the Division of Bio-statistics, Department of Public Health Sciences, University of Miami Miller School of Medicine, Miami, FL, USA. He obtained his PhD degree in the Department of Computer Science & Engineering (C.S.E.) from Jadavpur University, Kolkata, India, in 2017, while his PhD work was carried out in Machine Intelligence Unit (MIU), Indian Statistical Institute (ISI), Kolkata, India.

Dr. Anjan Bandyopadhyay is currently working as an assistant professor in the Department of School of Computer Science & Engineering at Kalinga Institute of Industrial Technology, Bhubaneswar, Odisha, India. He completed his PhD in the Department of Computer Science & Engineering from the National Institute of Technology, Durgapur, West Bengal, India. He has co-authored more than 30 research publications in various peer-reviewed international journals, conferences, and book chapters. He attended many national and international conferences in India and abroad. His research domains include cloud computing, fog computing, and algorithmic game theory.

Dr. Ruifeng Hu is currently working as a research fellow at Brigham and Women's Hospital, Harvard Medical School, Boston, MA, USA. Previously, he worked as a postdoctoral fellow in the Center of Precision Health, Department of School of Biomedical Informatics, University of Texas Health Science Center at Houston, TX, USA, for three years (2018–2021). His research is focused on the application of multivariate statistics, machine learning, and deep learning to omics data, and bioinformatics data.

Dr. Pawan Kumar Singh is currently an assistant professor in the Department of Information Technology at Jadavpur University, Kolkata, West Bengal, India, as of 2019. He obtained his PhD degree from Jadavpur University in 2018. Previously, he served as assistant professor in Department of Computer Science and Engineering, Techno India-Batanagar, West Bengal, India (2018), and Calcutta Institute of Technology (CIT), Uluberia, Howrah, West Bengal, India (2019). He has co-authored more than 100 research papers in various peer-reviewed international journals, conferences, and books. His research areas include pattern recognition, computer vision, handwriting recognition, machine learning, and artificial intelligence.

Prof. Soumadip Ghosh is currently serving as a professor in the Department of Computer Science & Engineering, Sister Nivedita University, Kolkata, India. He obtained his PhD from the University of Kalyani, West Bengal, India, in 2017. He has more than 18 years of teaching and research experience. He has co-authored more than 30 research papers, and his research domains are data mining, machine learning, and deep learning.

Dr. Sujata Swain is an assistant professor in the School of Computer Engineering at the Kalinga Institute of Industrial Technology, Odisha, India. She obtained her PhD and MTech in the Department of Computer Science and Engineering from Indian Institute of Technology (IIT) Roorkee, India. Previously, she had worked as an assistant professor at IIMT Engineering College, Meerut Galgotia's College of Engineering and Technology, Greater Noida, India. She has more than 40 publications, and her research interests include web service composition, service-oriented computing, and pervasive computing.

Contributors

Devansh Adwani
School of Computer Science and Engineering
Kalinga Institute of Industrial Technology
Bhubaneshwar, India

Chaithanya B N
Department of Computer Science and Engineering
GITAM School of Technology
Bangalore, India

Simandhar Kumar Baid
School of Computer Science and Engineering
Kalinga Institute of Industrial Technology
Bhubaneswar, India

Anjan Bandyopadhyay
School of Computer Engineering
Kalinga Institute of Industrial Technology
Bhubaneswar, India

Partha Sarathy Banerjee
Department of Computer Science and Engineering
Jaypee University of Engineering and Technology
Guna, India

Debajyoty Banik
School of Computer Science and Engineering
Kalinga Institute of Industrial Technology
Bhubaneshwar, India

Roshan Chatei
School of Computer Engineering
Kalinga Institute of Industrial Technology
Bhubaneswar, India

Parijat Chatterjee
School of Computer Science and Engineering
Kalinga Institute of Industrial Technology
Bhubaneshwar, India

Rudrashish Das
School of Computer Engineering
Kalinga Institute of Industrial Technology
Bhubaneswar, India

Geetha K
Department of Computer Science and Engineering
GITAM School of Technology
Bangalore, India

Bindu Madavi K P
Department of Computer Science and Engineering
GITAM School of Technology
Bangalore, India

Murali Kalipindi
Department of Artificial intelligence and Machine learning
Vijaya Institute of Technology for Women
Vijayawada, India

Jyotirmoy Karmakar
School of Computer Science and Engineering
Kalinga Institute of Industrial Technology
Bhubaneshwar, India

Aryan Kaushal
School of Computer Science and Engineering
Kalinga Institute of Industrial Technology
Bhubaneshwar, India

Rina Kumari
School of Computer Engineering, Kalinga Institute of Industrial Technology
Kalinga Institute of Industrial Technology
Bhubaneswar, India

G. Aloy Anuja Mary
Department of Electronics and Communication Engineering
Vel Tech Rangarajan Dr. Sagunthala R&D Institute of Science and Technology
Chennai, India

Shivansh Mishra
School of Computer Science and Engineering
Kalinga Institute of Industrial Technology
Bhubaneshwar, India

Megha Motta
Department of Electronics and Communication Engineering
Jaypee University of Engineering and Technology
Guna, India

Suchetan Mukherjee
School of Computer Science and Engineering
Kalinga Institute of Industrial Technology
Bhubaneshwar, India

Aditya Shankar Pandey
School of Computer Science and Engineering
Kalinga Institute of Industrial Technology
Bhubaneswar, India

Ritika Pandeya
School of Computer Engineering
Kalinga Institute of Industrial Technology
Bhubaneswar, India

Renuka R. Patil
Department of Computer Science and Engineering
GITAM School of Technology
Bangalore, India

Kartikeya Raj
School of Computer Science and Engineering
Kalinga Institute of Industrial Technology
Bhubaneshwar, India

Debarghya Roy
School of Computer Science and Engineering
Kalinga Institute of Industrial Technology
Bhubaneshwar, India

Bhaswati Sahoo
School of Computer Science and Engineering
Kalinga Institute of Industrial Technology
Bhubaneshwar, India

K. Aanandha Saravanan
Department of Electronics and Communication Engineering
Vel Tech Rangarajan Dr. Sagunthala R&D Institute of Science and Technology
Chennai, India

B. Sathyasri
Department of Electronics and Communication Engineering
Vel Tech Rangarajan Dr. Sagunthala R&D Institute of Science and Technology
Chennai, India

Deepak Sharma
Department of Electronics and Communication Engineering
Jaypee University of Engineering and Technology
Guna, India

Priyanshu Singh
School of Computer Engineering
Kalinga Institute of Industrial Technology
Bhubaneswar, India

Roshan Singh
School of Computer Engineering
Kalinga Institute of Industrial Technology
Bhubaneswar, India

Parivesh Srivastava
School of Computer Science and Engineering
Kalinga Institute of Industrial Technology
Bhubaneswar, India

Kartick Sutradhar
Department of Computer Science and Engineering
IIIT Sri City
Chittor, India

Sujata Swain
School of Computer Engineering
Kalinga Institute of Industrial Technology
Bhubaneswar, India

Sapthak Mohajon Turjya
School of Computer Science and Engineering
Kalinga Institute of Industrial Technology
Bhubaneshwar, India

Priyanka Venkatesh
Department of Computer Science and Engineering
Presidency University
Bangalore, India

Ranjitha Venkatesh
Department of Computer Science and Engineering
GITAM School of Technology
Bangalore, India

Vamsidhar Yendapalli
Department of Computer Science and Engineering
GITAM School of Technology
Bangalore, India

Introduction

The term "metaverse" refers to a virtual, interconnected, and immersive digital universe that encompasses a multitude of virtual reality (VR), augmented reality (AR), and 3D environments. It is a concept that has gained significant attention in recent years as technology has advanced, bringing us closer to the realization of a fully realized metaverse.

Interconnected Worlds: In the metaverse, various virtual worlds, social networks, and digital experiences are interconnected, allowing for seamless movement and interaction across different platforms and environments. It aims to break down the barriers between different online spaces.

Immersive Technologies: The metaverse is made possible through cutting-edge technologies such as VR, AR, mixed reality (MR), and other immersive technologies. These technologies create more realistic and immersive experiences, blurring the lines between the digital and physical worlds.

Social Interaction: One of the central aspects of the metaverse is social interaction. Users can engage with each other in virtual spaces, just like they would in the real-world. This can include chatting, gaming, attending events, working, or even building and creating within this digital universe.

Economic Ecosystem: The metaverse has the potential to develop a robust economic ecosystem. Users can buy, sell, and trade digital assets, such as virtual real-estate, clothing, art, and more. It introduces new opportunities for creators, entrepreneurs, and businesses.

Decentralization: Some proponents of the metaverse envision it as a decentralized space where no single company or entity controls everything. This could involve the use of blockchain technology to ensure transparency, security, and ownership of digital assets.

Challenges and Concerns: As the metaverse develops, there are various challenges and concerns to address, including privacy, security, content moderation, and the potential for addiction. There are also questions about the digital divide and who gets access to the metaverse.

Evolution of the Internet: Some see the metaverse as the next step in the evolution of the internet, moving beyond the 2D web to a 3D and highly interactive environment.

Companies like Facebook (now Meta), Google, and a range of startups are investing heavily in developing the metaverse, making it an exciting and rapidly evolving concept. The metaverse holds the promise of revolutionizing how we interact, work, play, and conduct business in the digital age, and its full potential is still unfolding.

In Chapter 1, we have introduced the concept of the metaverse.

The concept of "metaverse" is considered to be the next technological big bang after the advent of the internet. Metaverse views the internet as an expansive, unified, perpetual, and shared kingdom. Even though the implementation of metaverse is still considered to be future-bound, because its underlying philosophy is destined to integrate technology and digitize every aspect of our daily lives, and to achieve such modernization, it is paramount to further improve technologies like extended reality (XR), artificial intelligence (AI), and 5G, which are still not up to the standard needed for the metaverse implementation. The motive of this paper is to describe the salient tech development phases through which the metaverse should pass to get implemented in a large-scale environment. Alongside this, we also look after the current tech implementations that have already been made and closely examine their limitations to determine their functional efficiency in the metaverse. Therefore, giving the notion that the metaverse is not any disjoint technology rather it is an intelligent consolidation of various technologies that are already prevailing or will be invented in the future which is analogous to the elements that has conjoined in an orderly proportion to form our mother Earth.

In Chapter 2, we have introduced the digital transformation in healthcare. The digital transformation in healthcare is accelerating, powered by the convergence of the metaverse, AI, the Internet of Things (IoT), and blockchain technology. This review paper provides a comprehensive analysis of these technologies' role in reshaping healthcare from an engineering perspective. We explore the fundamentals of these technologies and their applications in healthcare, with a particular focus on system design, robotics, material science, and biomechanics. We present real-world examples highlighting the practical applications of these technologies, followed by an examination of the technical, ethical, and regulatory challenges they pose. Lastly, we discuss future prospects and the evolving role of engineers in this rapidly changing domain. Our review suggests that the digital healthcare journey, though laden with challenges, promises unprecedented benefits in terms of patient care and system optimization. Metaverse is now a connection between the real-world with digital world, where VR and AR are basically involved.

The metaverse represents an evolution of the internet into fully immersive and interconnected virtual environments. This review paper examines the technological roadmap and key applications of the metaverse across sectors including healthcare, education, banking, and blockchain technology. A conceptual analysis reveals the metaverse integrates identity modeling, decentralized technology, and social computing to create a VR experience. Core technologies facilitating the metaverse include VR, AR, MR, and XR. VR offers full immersion via headsets disengaging users from physical reality. AR overlays digital information onto the real-world through mobile devices. MR blends physical and virtual worlds for enhanced interactivity. XR encompasses the spectrum from VR to AR. Diverse applications in healthcare highlight metaverse potential in surgical training, clinical care, pain management, chronic disease education, and telemedicine. Educational implementations feature immersive learning, collaboration, and skill development. The metaverse provides the banking industry opportunities for enhanced customer communication, digital identity services, transparent lending, and contribution to carbon neutrality goals. Blockchain technology enables secure transactions, verifiable identities, and decentralization within the metaverse. However, challenges remain regarding accessible design, privacy, and responsible development. In conclusion, the metaverse signifies a major evolution in human–computer interaction, with significant implications across industrial, educational, medical, and financial domains that warrant further research.

This paper provides a comprehensive exploration of the convergence of Digital Twins (DT) and the metaverse, central technological forces in contemporary discourse. We delve into their foundational concepts, designs, and underlying architectures while shedding light on the inherent challenges they face. The analysis transitions to the real-world operations of the DT-powered metaverse, emphasizing its transformative potential in sectors like healthcare and heritage conservation. Through this investigation, readers gain insights into the complexities, operational mechanics, and far-reaching impacts of these digital innovations.

This research paper explores the idea of data driven smart cities in the metaverse. The paper examines the potential impact of the metaverse on physical cities and urban development, including the role of virtual currencies and other digital assets within the metaverse and the potential for these assets to exacerbate existing economic and social inequalities. The paper also discusses the potential for the metaverse to disrupt existing power structures and create new forms of social and political organization, along with the ethical considerations surrounding these possibilities. Furthermore, the paper examines the ethical implications of the metaverse, including the impact on social interactions and relationships, data use, ownership, and protection, and the potential for reshaping traditional power structures and its implications for democracy and social justice. Lastly, the paper discusses the role of policymakers in addressing the ethical implications of the metaverse. Overall, this paper aims to contribute to the ongoing discourse on the impact of emerging technologies on society and provides insights for policymakers and stakeholders to address the ethical challenges posed by the metaverse.

The developing metaverse provides a unique structure for complex interactions between different stakeholders, including customers, companies, and governments. To analyze these intricate relationships and direct the metaverse's future growth, this study uses Game Theory as a strong analytical framework. The study highlights four crucial areas where game theory can be used as both an analytical tool and a design guide. First, it makes it possible to comprehend how choices are made in the metaverse's virtual economies, encouraging ethical business practices and competitive growth. Second, Game Theory contributes to the development of more cohesive virtual societies by offering essential insights into the social dynamics of online communities. Third, the study looks at the regulatory implications of game theory and offers reasonable methods for governance that support both consumer protection and innovation. Finally, ethical issues are investigated, concentrating on just resource distribution and players' moral obligations. In conclusion, game theory provides a thorough, multidisciplinary lens through which to protect and optimize the metaverse's changing topographies. Now, we come to our new concept, which focuses on game theoratic auction-based approach used in the metaverse platform. Auction-based virtual resource allocation in the metaverse is a method for effectively allocating virtual resources inside a virtual environment. This approach utilizes auctions to determine the allocation of resources, with the goal of maximizing resource utilization and minimizing resource waste. The auction process involves multiple bidders competing for the same resources, with the winner being the highest bidder. Users can submit many requests at any given time using a pay-as-you-go basis, but only one request can ever be fulfilled. This situation is regarded as a "multi-requirement, single-minded scenario." We show how resource suppliers can increase social welfare while preserving the system's integrity. To quickly obtain the allocation outcome and increase the social welfare of the virtual resource provider, we propose a heuristic solution to the resource allocation problem. Overall, the allocation and use of virtual resources in the future may be completely changed via auction-based metaverse resource allocation. Next, we have focused on the application of metaverse in healthcare.

The application of metaverse is rapidly increasing in the field of healthcare. It is creating an impact by offering the best and optimal solutions that can improve patient care, medical training, and research work. Communication with patients with the help of visiting them virtually through metaverse reduces the need for physical appointments. Metaverse enables the interaction between patients and doctors in such a way that it becomes easier for the patient to discuss symptoms, diagnoses and even receive prescriptions. This method can be most effective in the field of remote or underserved areas. Medical professionals and students can use the platform of metaverse to practice surgical procedures, diagnose new diseases or infections, and refine their skills through such simulations. Complex surgeries can be performed in such virtual environments, which would reduce the risk of failure of any surgical procedure on real-patients. Such technology reduces risks, enhances training procedures, and also accelerates the skill development of surgeons. Another most recent application of metaverse is used to find out the fake medicine detection using blokchain technology. Blockchain Technology also implemented in metaverse. Counterfeit medicines present a dire threat to global health, with far-reaching implications for patient safety and public well-being. This research delves into the multifaceted impact of fake medicines and proposes a groundbreaking approach to their detection using blockchain technology within the metaverse, specifically on the Polygon blockchain. Counterfeit drugs not only jeopardize individual health but also undermine healthcare systems, erode public trust, and fuel criminal enterprises. Leveraging the attributes of the Polygon blockchain, such as scalability and low transaction fees, we present a robust framework for tracing and verifying the authenticity of pharmaceutical products within the metaverse. Through the tokenization of medicines as non-fungible tokens (NFTs) and the utilization of smart contracts, a secure and decentralized system is established, empowering users to authenticate medicines and participate in a transparent marketplace. The integration of QR codes or NFC tags enhances user engagement, while real-time tracking and data transparency foster informed decision-making. By exploring the dynamic synergy of blockchain technology, the metaverse, and the Polygon blockchain, this research contributes a comprehensive solution to the urgent challenge of counterfeit medicine, paving the way for a safer and more trustworthy pharmaceutical ecosystem.

In the next chapter, we have explored the integration of the Quantum Internet of Things (QIoT) with smart healthcare services within the metaverse, presenting a paradigm shift in healthcare delivery. By fusing quantum technologies with VR environments, a novel ecosystem emerges where quantum-enabled sensors, secure communication, and advanced computing reshape healthcare interactions. Quantum-enhanced sensors monitor physiological data with unmatched precision, seamlessly transmitting real-time information to the metaverse. Quantum cryptography ensures the security and privacy of sensitive health data exchanged within virtual consultations, paving the way for remote telemedicine and personalized diagnostics. Quantum computing optimizes complex simulations, expediting drug discovery and treatment simulations. However, challenges like quantum hardware stability and user interface design must be navigated. This chapter highlights the transformative potential of integrating QIoT with the metaverse in healthcare, offering innovative solutions that redefine patient care, diagnostics, and medical research. In the next chapter, we have focused on Health care systems. As healthcare systems become increasingly digitized, the need for secure and efficient data management becomes paramount. This chapter explores the fusion of two cutting-edge technologies: quantum computing and blockchain, to address the challenges of data security, privacy, and interoperability in the healthcare sector. By leveraging the unique properties of quantum mechanics, quantum blockchain offers the potential to revolutionize healthcare data management, enabling secure sharing, analysis, and storage of sensitive medical information. This chapter delves into the fundamental concepts of quantum computing, blockchain technology, and their convergence in healthcare. It presents real-world use cases, benefits, and challenges, and discusses the ethical and regulatory considerations surrounding the adoption of this innovative approach. Nowadays, AI has become a challenging area in the last decade. Nowadays, AI has left its footprints in almost all sectors of human life. For example, personal assistants such as Google Assistant, Siri, Alexa, etc., in search engines, automated vehicles, new advanced computer gaming sectors, and many more. In recent times, AI has also stepped into the medical diagnostics sector, which helps to provide better care and speedy recovery of patients leading to improved complete health care with much greater accuracy. Natural Language Processing (NLP) is known as a sub-branch of AI that helps the machine to analyze and compute the natural language that we humans speak. It helps to establish the communication between the machine and humans. With this ability of NLP, it is now used for the evaluation of medical test reports, e.g., radiological images that may include X-rays, MRIs, CT scans, blood reports, pathology slides, or Electronic Medical Records (EMRs). It enhances the capabilities of the physicians by providing aid in the diagnosis and treatment of the patient. This chapter helps to demonstrate the basic model of NLP working in the healthcare domain. It also includes different applications of NLP that contribute to the healthcare sector by increasing the accuracy of medical tests. NLP in healthcare not only provides support for healthcare professionals but also provides a platform for patients and their families to understand complex EMRs easily and make them much more vigilant toward their illnesses.

1 Technologies That Will Fuel the Future Metaverse and Its Potential Implementation in the Healthcare System

Sapthak Mohajon Turjya
School of Computer Science and Engineering, Kalinga Institute of Industrial Technology, Bhubaneshwar, India

Aditya Shankar Pandey
School of Computer Science and Engineering, Kalinga Institute of Industrial Technology, Bhubaneshwar, India

Anjan Bandyopadhyay
School of Computer Science and Engineering, Kalinga Institute of Industrial Technology, Bhubaneshwar, India

Sujata Swain
School of Computer Science and Engineering, Kalinga Institute of Industrial Technology, Bhubaneshwar, India

Debajyoty Banik
School of Computer Science and Engineering, Kalinga Institute of Industrial Technology, Bhubaneshwar, India

1.1 INTRODUCTION

The word "Metaverse" is formed by conjoining Meta which implies going beyond, with the word universe which refers to a theoretical environment built in connection with the real world.

In one sense, Metaverse can be regarded as a synthetic or computer-generated world that transcends or is much more ahead of the world that we currently live in. The concept of Metaverse was first introduced by an American writer named Neal Stephenson in his science fiction novel called the "Snow Crash" [1].

DOI: 10.1201/9781003449256-1

Stephenson depicted Metaverse as an enormous virtual environment where people or the users using it communicate or interact with each other with the help of digitalized or virtually created entities named the "Avatars."

Since its first introduction, many supporting concepts have been used to light up the existing concept of Metaverse with much better clarity to explain it further, such as an omniverse, lifelogging, and spatial internet. We will consider Metaverse as a virtual world manufactured by combining real and digitalized entities, empowered with high-speed internet, along with the capabilities of extended reality (XR). As per Milgram and Kishino's reality-virtuality continuum [2], XR blends real and computerized entities to various extents and the results of which are virtual reality (VR), augmented reality (AR), and mixed reality (MR). In the novel "Snow Crash," Metaverse was depicted as a duality between the physical and the computer-generated world. The entities present in the Metaverse to represent us and interact with other users are termed Avatars. Analogous to the real world where we carry out our daily life activities, Avatars will achieve the same in the virtual world driven by artificial intelligence (AI) and the user's coordination.

To attain such dualism, it is inevitable for the Metaverse to go through these three phases of development: (1) digital twins (Figure 1.1), (2) digital natives, and lastly (3) the existence of both physical reality and VR (Figure 1.2). Digital twins are massive-scale, highly precise computer-aided models replicated from real-life objects to implement them in the virtual world. Alongside duplicating, it also possesses all the properties sustained by real-life entities such as temperature and object motion. Both the virtual and physical twins are conjoined by the help of data or we can say data are the driving force of the virtual world. After creating a virtual copy of the physical world, the next big step is to leap forward in native content creation. The Avatars perhaps can be regarded as the content creators that is the entities depending upon which the scenarios in the virtual world are stimulated. These Avatars are complementary objects to that of the entities existing in the physical world. To make the ecosystem formed in the Metaverse even more attached or more relevant, laws

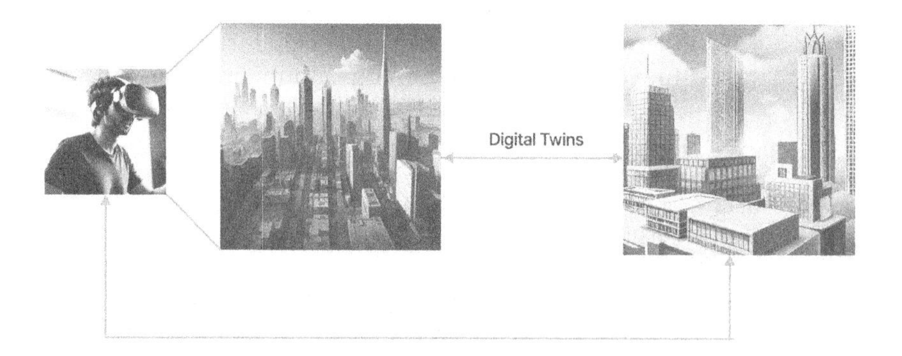

FIGURE 1.1 The person sees the real world through a VR gadget and interacts with the Metaverse.

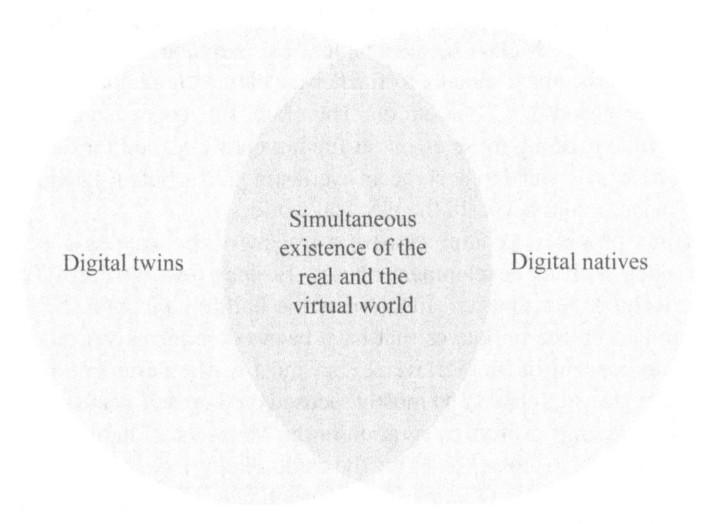

FIGURE 1.2 Digital twins-digital natives continuum.

and regulations, culture, and norms present in the physical world can be introduced [3]. Even though the research that has been put forward regarding the development of the Metaverse is still in the primary stage, mostly concentrating on the input technologies to gather the input data from the users and content creation [4], but in the ultimate phase it is expected that the Metaverse will become self-reliant and operate with the physical world as independently as possible.

Apart from these, the Avatars representing the humans in the real world can be configured to operate in various situations (or various virtual worlds) or to communicate with other coinciding Avatars alike to that are present in the physical world. Thus, following from this, the Metaverse can be enabled to operate between different virtual worlds allowing the users from the input side to create their worlds or scenarios and distribute content among these. To connect the users with its Avatars and allow them to operate in the virtual platform, ergonomic head-mounted wearable headsets can be introduced, which are already existing and used in many VR games like Half-Life: Alzx and also while watching 3D movies. As per the description given earlier regarding the diverse computer-aided reality, it is not surprising for someone to argue that we already lead our life in the Metaverse. Even though it is slightly true, but if compare it with the standards needed for creating the Metaverse, then we can identify the flaw in the statement earlier. For example, social networking sites allow their users to create their own content but they are all restricted to only creating posts, comments, text, and videos. Even though the essence of Metaverse can be more deeply experienced in video games like the Call of Duty where the real-world scenarios are depicted in much greater detail, but still, these video games cannot exhibit the concept of interoperability, which is an essential requirement for the Metaverse creation.

Alongside technologies like social networks, internet, and the virtual domain in order to understand the Metaverse even better, it is paramount to have a firm grasp and lay down improvement models to further develop technologies like AI, blockchain, computer vision (CV), and so on. Therefore, this chapter takes a vital step forward toward explaining these essential fundamentals needed for developing the Metaverse which is destined to become an everlasting, distributed, synchronous, and virtual workspace conjoined with the physical world.

This chapter provides a comprehensive overview of the existing concept of the Metaverse along with its development phases. Besides this, we deeply focused to explain the technologies that will function as the building block of the Metaverse. We also summed up the initiatives that have been taken in the various tech fields to light up the concept of the Metaverse, but most of the developments that have been already made are isolated and mostly focused on their self-centric development rather than considering a direct integration in the Metaverse. That is why, we think it much very essential to closely examine the compatibility between the technologies that are already prevailing in the market to consolidate into the Metaverse, but this is beyond the scope of this chapter, and we keep it as a matter of further research. In this chapter, we have also made an effort to identify the various loopholes in the mentioned technologies focusing mostly on the functional and the security parts. Again how these loopholes will be filled or what will be the proposed development model, we are dedicating this to a wider research community.

Alongside technological components, the content that will be previewed in the Metaverse is also an integral part of the whole Metaverse discussion. Even though we have briefly discussed this highly extensive field, but deeper description has been discarded to give more emphasis on the technology section, thereby we keep this matter reserved for further research.

1.2 FRAMEWORK

As the concept of Metaverse is very much diverse and sophisticated, this section's goal is to briefly describe the connection between the various key ingredients of the Metaverse categorized under two fields: technologies and ecosystems (Figure 1.3).

Beginning from the technology part, users interested can handle the Metaverse with the aid of XR and user interactivity procedures. Alongside this, technologies like AI, blockchain, CV, and Internet of Things (IoT) can function in correspondence with the user in the physical world to accomplish a plethora of handful tasks in the Metaverse. The function or application of edge computing in the Metaverse is to upgrade the workability of certain applications which are latency prone and bandwidth-consuming. It achieves this by utilizing data from the regional data repository as previously processed data operable in the edge gadgets. On the other hand, cloud computing is well-reputed for its extensively expandable space capacity and processing strength. Thus, it can be concluded that both edge- and cloud-based services can immensely increase the application efficiency and therefore make a notable difference in the user experience with the further addition of high-speed internet network (5G, 6G, or more advanced) acting as the base for CV, AI, and IoT with top-notch viable hardware.

Technologies that will act as the building blocks of the Metaverse	Components that makes up the Metaverse ecosystem
• Artificial intelligence • Blockchain • Computer vision • Network • Edge computing • User interactivity • Extended reality • Internet of things	• Avatar • Content development • Virtual economy • Social acceptance • Security • Privacy • Trust • Accountability

FIGURE 1.3 Technologies that make up the Metaverse, and the contents of the Metaverse ecosystem.

Ecosystem depicts the contents, which are counterparts or digital twins of the objects sustaining in the physical world. Users residing in the real world can interact with their virtual counterparts or Avatars via XR and user interactivity techniques for performing different tasks like content creation or modifying the ones that have been already created. Alongside these, various real-life aspects like social acceptableness, virtual economy, privacy, security, accountability, and trust can also be contemplated in the Metaverse with other aspects such as social norms, customs, laws, and regulations, which should be synchronized with the contents created or with the ones to be developed. For example, the creation of a virtual economy should be protected with great security and it should be trustworthy and acknowledged by the other Avatars present in the Metaverse. The most important of all, the data security and privacy of all the users in the Metaverse should be taken into account with great care.

Figure 1.4 is a pictorial representation about how we blend all the technologies to make them functional in the Metaverse. In the next section, we will go through the salient technologies which are required for turning our concept of the Metaverse into reality. These are (1) XR (VR, AR, MR), (2) user interactivity, (3) Internet of Things, (4) AI, (5) CV, (6) blockchain, (7) edge computing, and (8) network.

1.3 EXTENDED REALITY

1.3.1 Virtual Reality

VR gives us the vision to unfold the exciting features of the computer-aided artificially created universe. The VR equipment already available in the market in the form of ergonomic wearable headsets and tactile controllers provides straightforward means for users to communicate or interact in the virtual world. Therefore, the users can establish a connection with their virtual counterparts and interact by its means following specific control methodologies. In the reality-virtuality continuum [2],

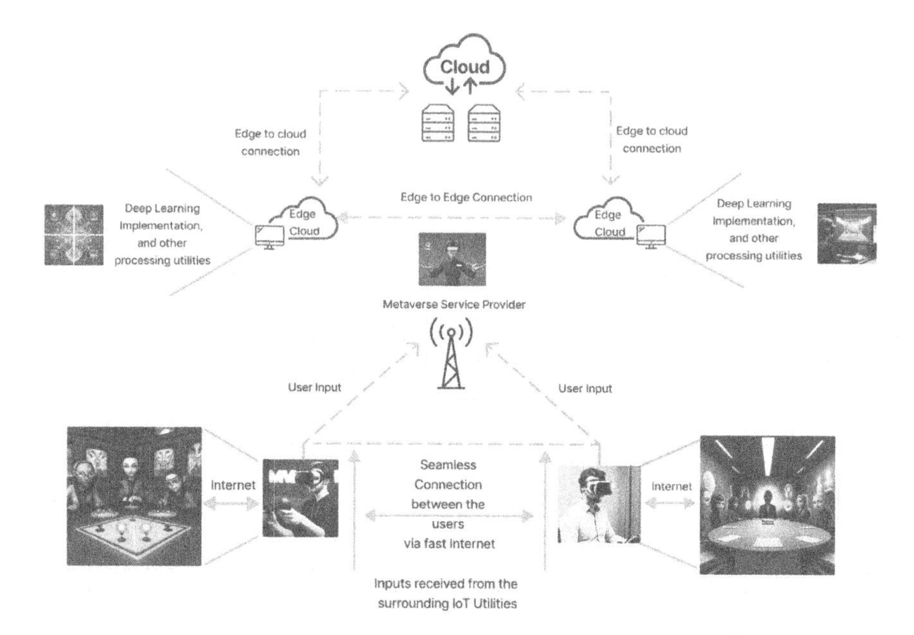

FIGURE 1.4 A simplified Metaverse model.

VR is situated very far away from reality, and because of this, we notice users using VR headsets become completely engrossed in the virtual world and therefore barrier themselves from the physical world. As mentioned earlier, Metaverse allows users to create their content in the virtual world; currently, such environments are already available in the market where the user can build and customize the content, for example, VR painting. Discovering the possible scopes through which the users can utilize the virtual world can be best attained by providing users a real-time virtual world/reality experience. For example, the user may want to change the size of the virtually created objects or might be interested in creating new objects from the ones that are currently existing or just be creating contents from the scratch. Alongside this, more than one user can be introduced into the virtual ecosystem and can combine and accomplish certain tasks simultaneously. This completely matches with prerequisites of the virtual ecosystem: that to have a distributive sight of space, a distributive sight of presence, a distributive sight of time, and a possible method to interact and transmit data or information with the other users [5]. Alongside these, it should be maintained that every user receives the same information as the others, and users should also be availed of the facility to communicate with each other simultaneously.

It is a huge challenge how a user will govern these virtually created entities and work in combination with other users sharing the same virtual ecosystem. Being in the last stage of the Metaverse creation, the next task would be allowing users sustaining in the virtual ecosystem to communicate and act accordingly with the physical world, and this would be feasible through the blending of AR and MR with VR. The heart of the Metaverse is constructed by composing multiple virtual workspaces with all the Avatars representing their physical counterparts and supporting all types of communication such as object-object, object-avatar, and so on. And all these preceding processes should function in such a manner that they all should be dynamic in nature and timed to perform in real time [6]. As of now, all the technologies that have been discussed to be implemented in the Metaverse are somehow present or in the development stage, but the main challenge would be merging these technologies together and synchronizing them to perform in real time. Because as we have seen earlier, the Metaverse is composed of the words Meta and Universe, and the universe we currently live in is very much dynamic in nature, thus achieving such dynamism without any latency, and keeping the other factors in mind would be a huge challenge to be accomplished in the future. In the next phase, we will be going through the other technologies needed to be implemented in the Metaverse.

1.3.2 Augmented Reality

The term AR differs from VR because it is the real-time processing and blending of real-world data using various computer interfaces. Usually, it consists of three phases: detection, tracking, and merging. The AR system uses various tracking and registration algorithms to accurately and efficiently track the movement and position of real-world users and objects. These algorithms use a combination of sensors and cameras to detect the location and orientation of entities in the environment and map virtual entities and environments to the real-world AR systems can also use AI and machine learning technology to enable smarter, more adaptive behavior in virtual worlds.

One needs to create engaging, intuitive AR experiences that are fun and easy to use for users of the virtual world. Potential solutions to address this problem include designing natural and intuitive AR interfaces and interactions, developing engaging and interactive educational content, and evaluating the effectiveness and usability of AR systems. This includes continuously improving the user experience.

Metaverse AR must ensure user privacy and security, especially when using AR systems in sensitive or personal contexts within virtual worlds. Appropriate security measures and protocols, such as encryption and authentication, should be implemented, and policies and guidelines should be developed for the use of AR in various domains within the Metaverse [7].

1.3.3 Mixed Reality

MR, also known as hybrid reality, is a term that refers to the combination of real and virtual environments and the interaction between them. It is a rapidly evolving field that has the potential to revolutionize the way we interact with the world and with each other.

By mixing real and virtual elements together, MR permits users to communicate with virtual entities and environments in a more natural and intuitive way, using their own physical movements and gestures. This could potentially enhance the realism and immersion of virtual experiences, making them more compelling and engaging for users.

AR/VR development frameworks are software libraries and tools that are specifically designed to speed up the development of AR and VR applications.

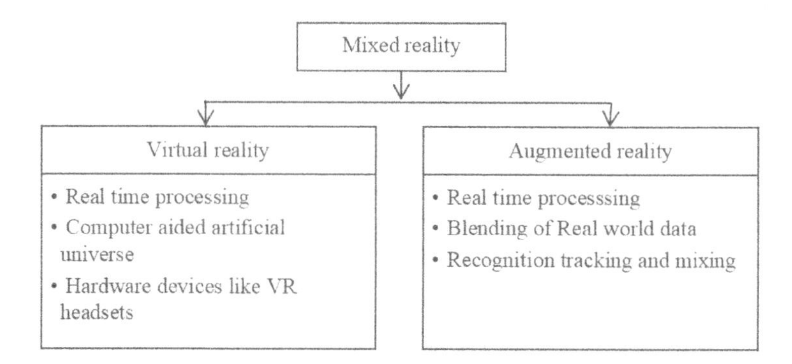

FIGURE 1.5 A visualization of mixed reality using augmented and virtual reality.

Some examples include ARCore, ARKit, and Vuforia. Haptic feedback technologies enable users to feel virtual entities and environments in a more immersive and realistic way. These technologies can be integrated into MR applications using software interfaces and APIs. NLP technologies are used to enable MR applications to understand and respond to natural language inputs from users. These technologies can be integrated into MR applications using software libraries and APIs.

MR, in the Metaverse, is a tool for enhancing communication and collaboration. By using MR technologies, people can work together and interact with each other in virtual spaces, regardless of their physical location. This could enable more efficient and effective collaboration and could potentially facilitate the creation of virtual communities and social networks within the Metaverse. MR could potentially be used to create virtual art galleries that permit users to experience and communicate with art in a more immersive and interactive way. By using MR technologies, users could see how artworks look and feel in the real world, and could potentially even create their own virtual artworks using MR tools.

Users could see and hear each other as if they were really there and could potentially even touch and feel each other using haptic feedback technologies. Some of the other potential applications of MR in the Metaverse could be virtual education, virtual tourism, virtual entertainment, and virtual training and simulation.

Overall, MR has the potential to play a significant role in the development and evolution of the Metaverse, and it is likely that new and innovative uses for MR in the Metaverse will emerge as the technology continues to evolve [8]. Figure 1.5 represents a visualization of MR using AR and VR.

1.4 THE AVAILABLE USER INTERACTIVITY DEVICES FOR IMPLEMENTATION IN THE METAVERSE

At the last stage of the Metaverse creation, the real world will be conjoined with its digital twin, which means all the users present in the physical world will be able to communicate with their virtual counterparts and also with the digitalized objects existent in the Metaverse and the MR in the real world. Alongside this, there would be an omnipresent connection established between the users and virtual entities existing throughout the Metaverse. So, we can say that the real and the virtual world can thus affect each other persistently.

Most of the existent VR simulators allow the users to control their Avatars with the aid of a mouse and keyboard, which is very much inefficient to replicate the real-world body movements expected from the Avatars in the Metaverse. Another drawback of this kind of interacting device is that the user needs to be stagnant (e.g., sitting) [9] while controlling their virtual counterparts, which is not at all suitable for mobile users. The possible scope to operate freehand promotes innateness due to the presence of barehanded actions [9], and such freehand reciprocity to interact can be attained with the help of CV procedures. Even though it is very challenging to dynamically interpret the freehand gestures in the real world with optimum precision and replicate the same in the virtual world, the most basic mid-atmosphere pointing needs the application of optimal computational utilities. Implementation of inadequate computational resources will cause a delay to the user's action and thus will decline the user experience. Shifting from the CV-based controlling mechanisms, vast research has broadly transformed the user input style to support the sophisticated user-controlling functionality, including optical, pyroelectric infrared, electromagnetic, capacitive, and IMU-driven user-controlling mechanisms. These types of user interactive mechanisms can efficiently apprehend the user activities and thereby interconnect with the virtual objects existing in the Metaverse.

In this section, we will go through some of the existing approaches put forward to transform the field of user interactivity techniques. For example, ActiTouch (Figure 1.6) uses a capacitive layer mounted on the user's forearm. The electrodes used in ActiTouch convert the user bearing it into a capacious input device. Therefore, by simple nudges, the user can communicate with the other Avatars available across various platforms in the Metaverse. Another approach [10] enhanced the collection of input instructions, where users can communicate with the help of interactive menus,

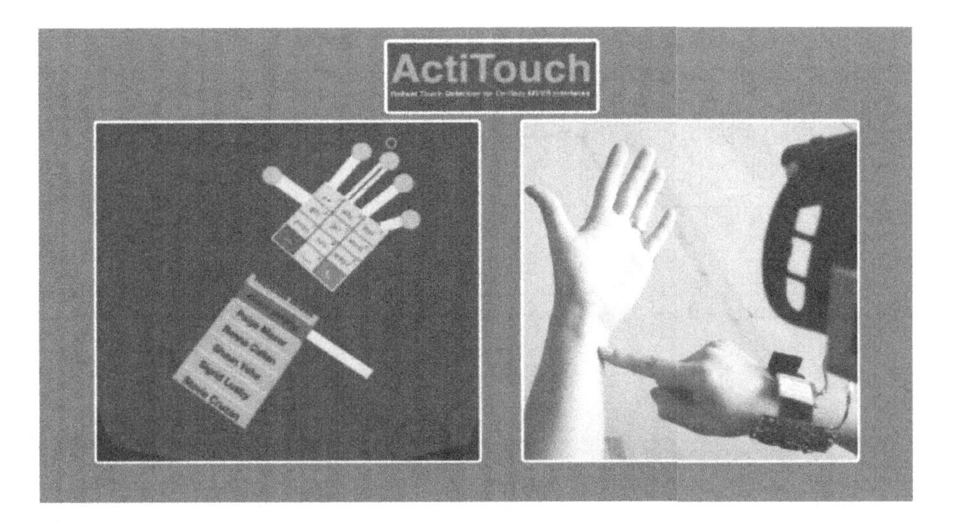

FIGURE 1.6 ActiTouch: Enabling touch detection for AR/VR interaction. The figure was inspired from Future Interfaces Group, YouTube.

icons, and so on, as if AR surfaced on the user's arms. This type of body-attached user interactivity technique can be used to stimulate remote touch sensations which will vastly boost the relevance of distant interactivity among users in the Metaverse. The most-present body-attached-users-interactivity techniques reflect a reduction in the equipment size ranging from the user's palm to their fingertips [11], which means the presence of an input device attached to the user's body is increasingly becoming less identifiable. However, focusing on different options does not mean that we are neglecting the scopes presented by CV. CV combined with the substitute models can be used to ensure innateness and solve complex user interactivity issues [9]. For example, CV employment can upgrade the working efficiency of the IMU sensors. The approaches proposed through CV can be used to calculate the distance between the user's hand with respect to the digitized entities, and the IMU sensors can be used to ensure the precise modification of the virtual entities.

Interestingly, tech giants like Google have taken steps forward and invested in this field. In the Jacquard project [12], Google strives to manufacture modernized woven at low costs and in high quantities. These smart wovens can be implemented with everyday wearables like jackets to receive real-time dynamic inputs. Alongside this, researchers of other well-reputed companies are striving to invent a cost-effective, precise, and minute user input device to implement it in the Metaverse. Figure 1.7 represents user interactivity and its various use cases in the Metaverse.

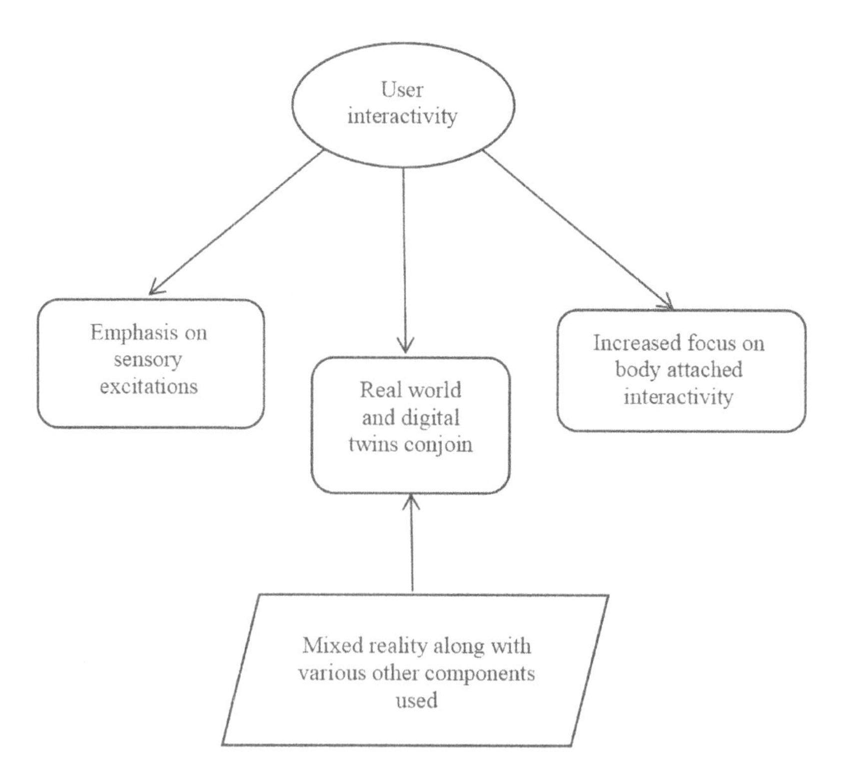

FIGURE 1.7 User interactivity and its various use cases in the Metaverse.

1.5 THE USE OF INTERNET OF THINGS IN THE METAVERSE

The number of IoT-implemented equipment integrated around the world is increasing at a fast rate (Figure 1.8), so in such scenarios, many tech researchers believe that the idea of merging IoT with VR/AR/MR will make it possible to implement multi-modal controlling mechanisms which will vastly improve the user experience, especially for the ones who are not expert handlers. The salient cause behind this is that it permits communication devices to merge the user's physical surroundings with realistic AR entities. The astounding developments put forward in the field of modern IoT systems has opened up a widespread possibility to implement infrastructures that will vastly modernize our everyday life. The small-scaled IoT equipment cannot support tactile interfaces for precise user communication. The virtual objects under the supervision of the XR can make up for the interaction equipment that was previously absent. That means users in mid-air can have a view of the XR contents [13]. Alongside this, some large-sized electrical equipment similar to that of robotic arms, due to form factors constraint, finds it advantageous to allow the users to operate the equipment from a distance, where XR plays the role of a mandatory controller. Therefore, users can barrier themselves from using controlling devices that are perceivable by tough, keeping in mind the fact that it is not feasible to integrate numerous controlling devices with respect to different IoT equipment.

Virtual ecosystems represent some eminent attributes for envisioning imperceptible objects and their related functioning, such as the WIFI and the user's data [14]. Besides this, AR also visually represents the data transmission via various IoT devices to the user in the physical world. Thereby, keeping the user alert regarding the data transmission between the user and the connected IoT equipment and giving the user the necessary control to regulate their data flow between the IoT devices through AR envisioning interfaces [14].

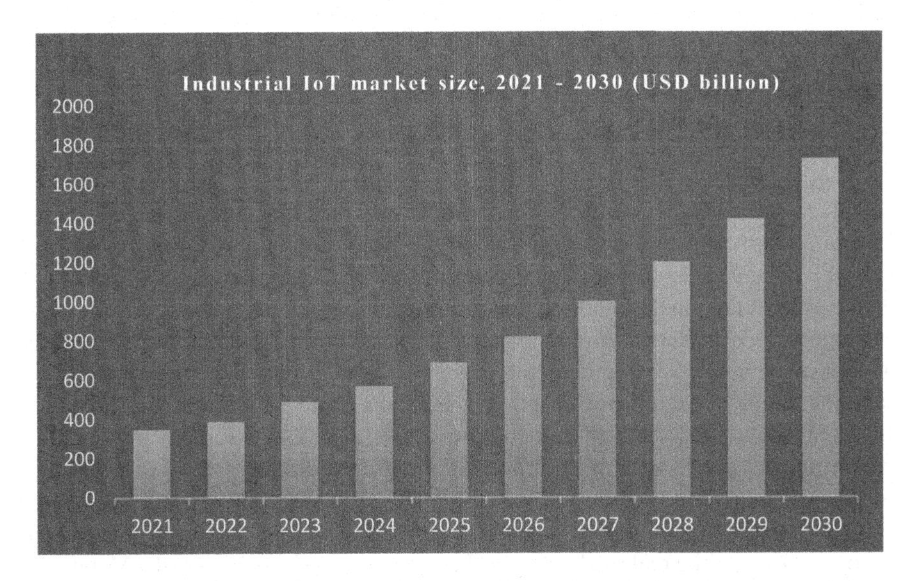

FIGURE 1.8 Industrial IoT market forecast. Graph inspired from precedence research.

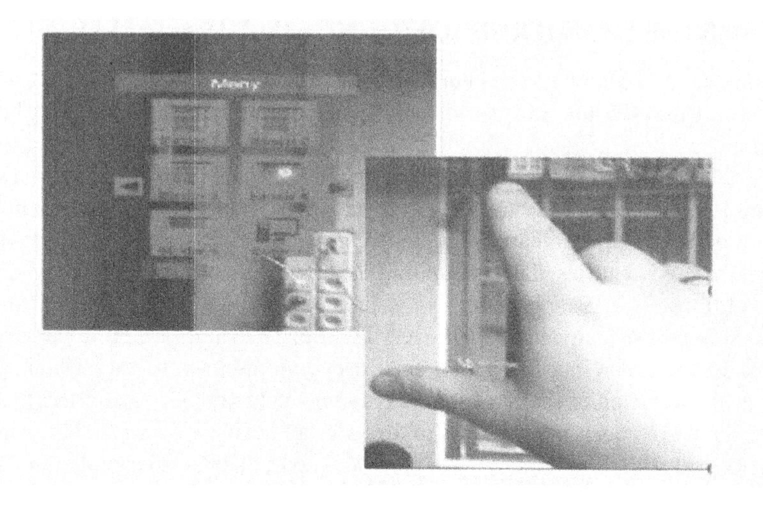

FIGURE 1.9 Mid-air icons previewing the XR contents perceivable by the users through natural gestures.

Mid-air interfaces permit users to regulate the integrated IoT devices with the help of normal controlling techniques [15] (Figure 1.9), such as scrolling, selecting, and so on. Vera [16] permits users to accomplish tasks by implementing an AR interface that will be hand-held; the AR system is connected to a robot, which executes tasks as per the instructions provided by the user. Alongside this, drones are very well-known IoT devices that are repeatedly implemented in XR. In multi-user systems scenarios, a drone can be operated by the cooperating users to accomplish specific tasks. Similarly, Pinpointfly [17] indicates a portable AR system that enables manipulation of the drone's movement and roadmaps via improved AR viewings.

1.6 THE IMPACT OF ARTIFICIAL INTELLIGENCE IN THE METAVERSE

AI has become the driving force of all the tech revolution that has been taking place since the beginning of the 21st century. The concept of AI first being introduced in the year 1956, but its implementation took upon a pace in the 21st century. Figure 1.10 represents the forecasted AI market, where a clear growth is observable in the current and the upcoming years which represents AI's increasing demand. AI was a concept but is now a destined faith or technology that aims at making machines/computers intelligent beings giving them the capacity to take the necessary decisions based on the scenarios to which they are exposed. In the present decade, it has attained up-to-the-minute execution in various fields whose application is very much relevant to human society.

AI in combination with technologies like AR/VR, networking, and blockchain can create virtual environments that will be dynamic, secure, and most importantly synonymous with the real world. As per the seven composing layers of the Metaverse (Figure 1.11), the application of AI is very much paramount to ensure the optimal

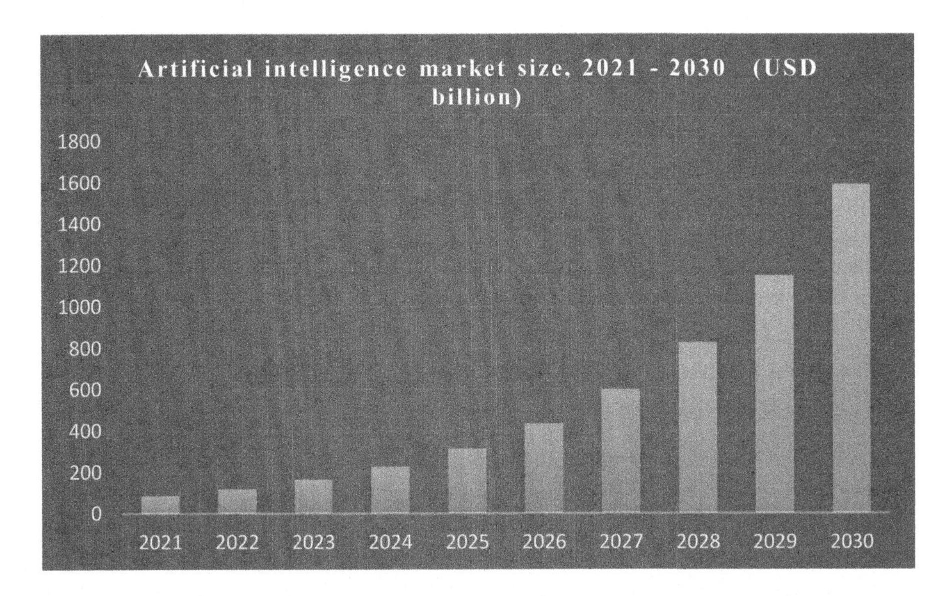

FIGURE 1.10 Artificial intelligence market forecast 2021–2030. Graph inspired from precedence research.

working efficiency and trustworthiness of the Metaverse platform. With the existing 5G and the future 6G networks, a plethora of advanced machine learning algorithms composed of supervised and reinforcement learning can be implemented to accomplish a handful of demanding tasks, which are attack avoidance, efficiency in spectrum monitoring, automated resource, allotment, detection of network failures, traffic resolving, and so on. Data obtained from the sensor-attached wearable equipment can be deeply analyzed to interpret and identify body locomotion and complicated actions on the basis of machine learning and deep learning models, thereby reflecting the same in the virtual environment with the help of the Avatars. Alongside these, AI integration in the Metaverse with optimal efficiency and computing speed

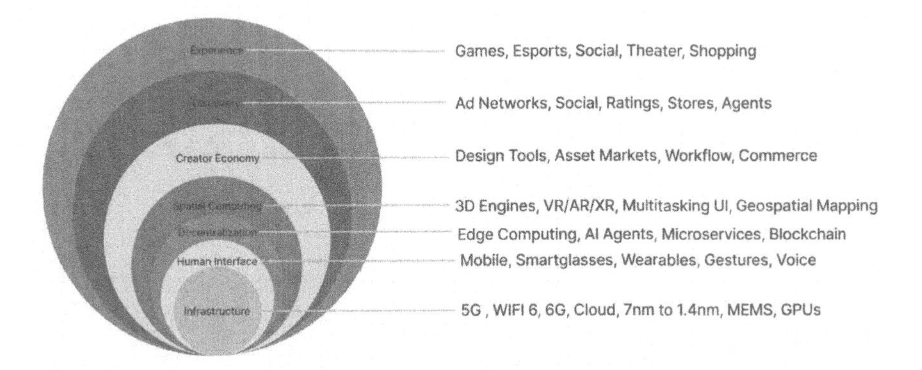

FIGURE 1.11 Seven composing layers of the Metaverse.

allows Avatars to perform real-world activities such as adjusting facial expressions based on the scenario, body locomotion, emotion analysis, speech differentiation, and sentiment analysis.

If we deeply analyze the concept of the Metaverse, then it is very much observable that the Metaverse is hugely data-driven, and data being the fundamental driving force of the Metaverse, it is very much obvious that the application of AI would be hugely demanding here. Metaverse being a dynamic entity would require real-time data processing as per the users' input or from the outputs produced, which is unthinkable without the application of AI or more specifically deep learning.

1.7 OBJECT TRACKING USING COMPUTER VISION IN THE METAVERSE

The role of CV in XR is very much vital for the future implementation of the Metaverse (Figure 1.12). The XR devices obtain optical data via video see-through layouts. These data are then processed and the outcomes are transmitted through smartphones and headsets. The application of CV utilizes this generated data by deeply analyzing them and thereby interpreting the obtained optical information as images and videos and thus extracting valuable insights to take the appropriate actions. Shortly, CV gives XR the cognizance to identify and differentiate between different visual information generated by different users and their surroundings in the real world. Simplistically, CV opens the possibility to develop a much more responsive, immersive, and precise virtual, augmented world.

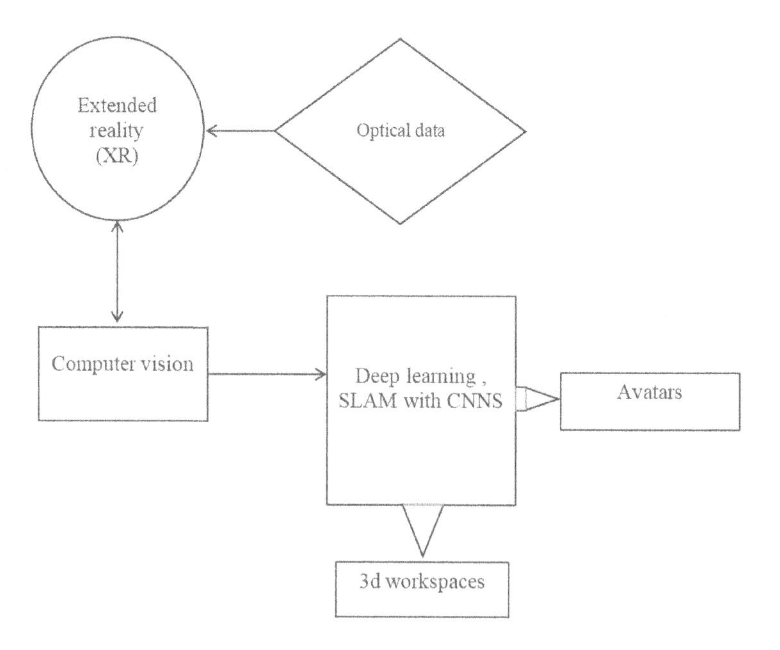

FIGURE 1.12 Prototypical working of computer vision laced with XR.

In the Metaverse, the users will get the opportunity to interact with their virtual counterparts (Avatars) and also simultaneously communicate with the physical world in which they are living. Prioritizing the key attributes of interoperability and digital twins, it is very much essential to have an in-depth cognizance regarding human activities and duplicate the same in the Metaverse by the Avatars. In the real world, we obtain structural data with our eyes and construct a three-dimensional shape of the real workspace in our brain, where we can identify the precise position of every nearby object. Likewise, the Metaverse also requires the three-dimensional reconstitution of the unmapped domain to which it is exposed, and also it needs to interpret object movement. Thereby to achieve this objective, CV techniques such as simultaneous localization and mapping (SLAM) can estimate device movement and rebuild a previously unknown environment [18]. A visual SLAM algorithm has to find simultaneous solutions to various demanding tasks, such as (1) real-time tracking, (2) unmapped spaces, (3) unmanageable camera, and (4) drifting issues. Various SLAM algorithms, including the ORB-SLAM series, and ORB-SLAM-v2 [19] seem to be compatible with AR systems, thus representing their relevancy in the Metaverse.

Visual SLAM algorithms are generally dependent upon these three fundamental steps, which include (1) feather extraction, (2) 2D frame mapping to the 3D point cloud, and (3) close-loop detection.

The first work for a lot of SLAM algorithms is to determine characteristic (feature) points and produce descriptors [18]. Conventional feature tracing procedures like scale-invariant feature transform [20] can identify and detail the local characteristics in images, but their drawback is that they are very much latency-prone to function in real time. That is why, the majority of AR systems rely on processing efficient tracing procedures, like feature-based identification [21], to check coinciding features in real time without the implementation of any GPU boosting. However, the recent implementation of convolutional neural networks in visual SLAM has seen a rise in the performance of self-governed driving with GPUs [22], but still, its implementation is very much difficult for resource-restricted mobile systems.

After tracing the characteristic points, the next task is to obtain the 3D coordinates related to point estimation by determining the mapping technique from the 2D frames [23]. When the camera captures and gives a new frame as an output, key points are then generated by the SLAM algorithm through estimation. The obtained points/coordinates are then mapped/plotted in the preceding frame to evaluate the ocular smoothness of the segment. Sometimes, the approximated camera pose is not that accurate. To solve these issues, SLAM algorithms like ORB-SLAM [19] combine more data to improve the existing camera pose by tracing more key points' similarities. New map coordinates/points are produced through triangulation of the coinciding key coordinates from the conjoined frames. This procedure packs up the two-dimensional locations of the key points and shifting and gyration between the frames.

Even though the visual SLAM algorithms have constructed a firm base for spatial interpretation, however, the Metaverse should be even more intelligent enough to interpret more complicated real-world scenarios. Thus, with such firm necessity, what we can expect is for the SLAM algorithms to become more superior in terms of mapping and processing efficiency for its implementation and successful creation of the Metaverse. Figure 1.12 represents the prototypical functioning of CV laced with XR.

1.8 BLOCKCHAIN: A DECENTRALIZED APPROACH FOR TRACKING VIRTUAL ASSETS IN THE METAVERSE

As per Satoshi Nakamoto, Blockchain is "A database of transactions shared by all nodes partaking in a system based on the Bitcoin protocol." A full copy of a currency's blockchain includes all the transactions ever made in that currency. With these data, one can figure out how much wealth owned by each address at any instance in history.

Using blockchain technology to create a decentralized and transparent platform for tracking and verifying virtual assets is an important aspect of Metaverse. In the Metaverse, virtual assets such as virtual currencies and virtual goods can be created to be traded and used to facilitate interactions in the virtual world. These assets may have real value and it is important to have a secure and transparent system to track and verify their ownership and transfer. The solution to this problem is offered by blockchain technology, which enables the creation of a decentralized and transparent platform for recording and verifying transactions. An important application of blockchain technology in the Metaverse is the implementation of smart contracts to facilitate interactions in the virtual world. These self-executing contracts with purchase clause between a vendor and the buyer are implemented by directly writing them into the lines of code. They can be implemented on the blockchain and can be enforced using cryptographic protocols.

VR content and assets may include virtual currencies, virtual goods, and virtual experiences and may have real-world value. Blockchain technology can be used to create a secure and transparent platform for tracking and verifying the ownership and transfer of these assets. VR marketplaces are platforms that allow users to sell and buy VR assets and VR services. Blockchain technology can be used to facilitate the exchange of these assets in VR marketplaces by enabling the safe and transparent transfer of virtual currencies and virtual goods. This could enable the creation of new VR experiences and business opportunities within the VR ecosystem [24, 25]. Figure 1.13 illustrates the application domains of blockchain in the Metaverse.

As stated earlier, Metaverse is the next technological big bang that we are to witness in the future. Table 1.1 represents the various characteristics of Web 1.0 and

TABLE 1.1
Characteristics of Web 1.0, Web 2.0, and Possible Web 3.0 (in the Future)

	Web 1.0	Web 2.0	Possible Web 3.0
Communication method	Read	Read-write	Read-write own
Communication channel	Static text	Responsive content	Virtual economies
Harassment and hate speech	52%	31%	16%
Infrastructure	On-premise	Cloud	IPFS and blockchain
Data control	Centralized	Centralized	Decentralized

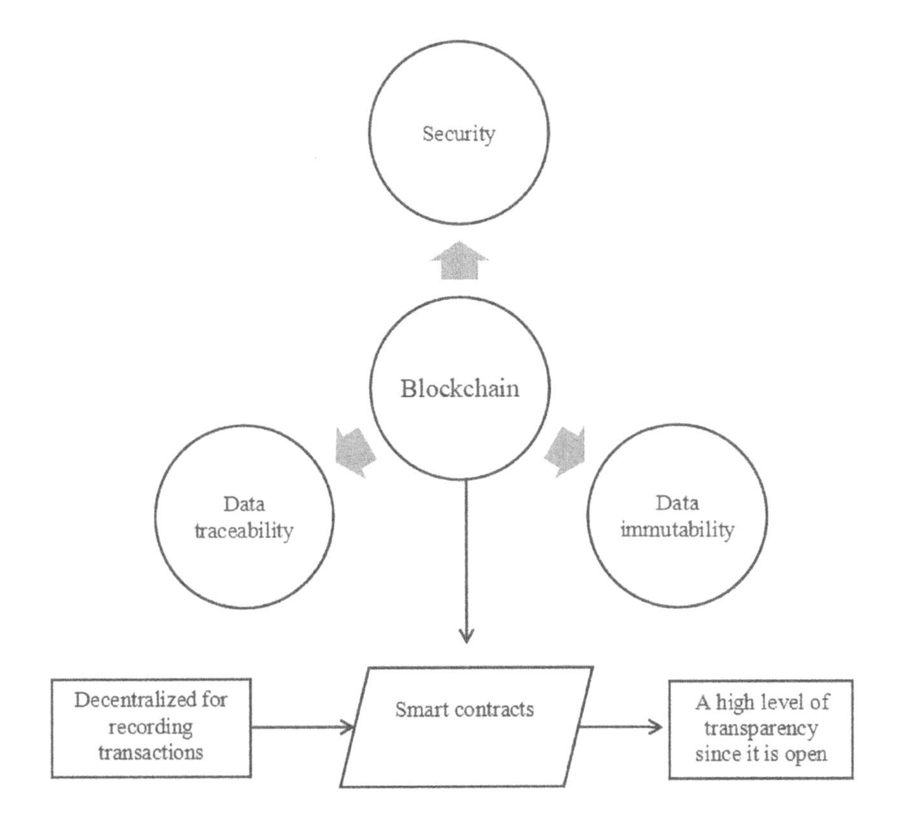

FIGURE 1.13 The scope of blockchain in the Metaverse.

Web 2.0 in terms of communication method, communication channel, organization, infrastructure, and data control. The next possible Web 3.0 (which is the Metaverse) is crafting a much greater possibility in terms of the communication channel, infrastructure, and data control merged with the advantages of the previous web versions.

1.9 EDGE COMPUTING TO INCREASE VR RESPONSIVENESS IN THE METAVERSE

According to many tech experts and researchers, the implementation of edge computing is very much vital for the creation of the Metaverse because it substantially reduces system delay in the virtual environment. For instance, tech giant Apple Inc. connected Mac with VR headsets for 360-degree VR displaying. Another example is, Facebook Oculus Quest 2 operates VR independently without any PC connection and the credit behind this goes to its strong Qualcomm Snapdragon XR2 processor. But still, it cannot fully utilize its potential in comparison to that of a high-configured computer, the result of which is a drop in frame rates and clarity in the VR stimulation. The implementation of edge systems boosts responsiveness, image/video resolution, and clarity and decreases delay to give a seamless VR experience.

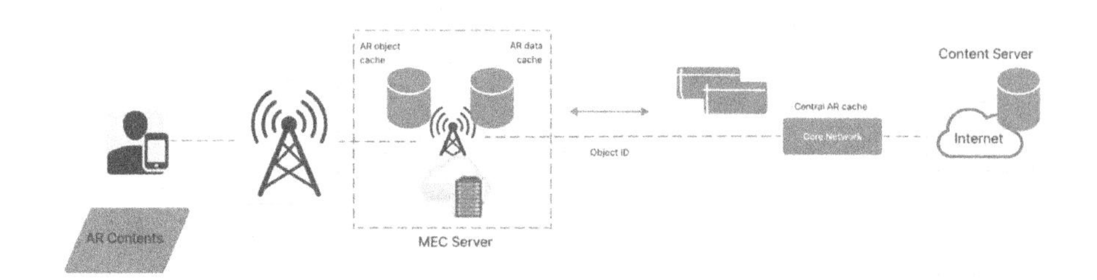

FIGURE 1.14 A MEC-based solution for AR applications [26].

To provide users with a seamless ubiquitous experience of the Metaverse, it is important to ensure an interruption-free outdoor internet network. The inauguration of the 5G and the future 6G is believed to vastly improve the outdoor internet network.

The multi-access edge computing (MEC) is believed to uplift the user experience in the Metaverse by implementing a universally standardized edge service that is one hop apart from the user's gadget connected by cellular networks, for example, AR glasses. The European Telecommunication Standards Institute put forward the MEC, which is a telecommunication-centered cloud-based edge computing model used for the implementation, functioning, and servicing of the edge services controlled by an ISP situated one hop apart from the base terminal [27]. MEC is very important while implementing the Metaverse outdoors because MEC successfully interprets the outside scene comprehensively and establishes an internal connection with the other user placed nearby. For example, the MEC systems equipped with 5G can handle the AR data of the surrounding users and provide the facility to communicate in real time. Figure 1.14 diagrammatically represents a MEC-based solution for AR applications.

In the present day, many companies investing in the Metaverse project have taken the initiative to implement MEC to boost the user experience. For example, Holoverse in association with TIM, Deutsche Telekom, and Telefonica have evaluated the 5G-fueled cloud-based edge network system for its implementation in various services with the help of the Metaverse. Niantic Inc., the company which designed AR games like PokemonGO and Harry Potter, visualizes creating an AR-like planet. It has also conjoined with other tech companies like Deutsche Telekom, Global Telecom, SoftBank Corp., EE, Orange, TELUS, Verizon, and Telstra envisioning to implement MEC for achieving an uplifted AR system. With the 5G technology and 6G emerging, the last mile lag will be more minimized. Thus, it can be concluded that MEC utilization will uplift its benefits in the Metaverse.

1.10 REQUIREMENTS FOR A SEAMLESS NETWORK CONNECTION IN THE METAVERSE

As most of the multimedia content in the Metaverse will be in high definition and dynamically timed for providing users with an immersive and realistic experience, thus it demands data transmission with very high bandwidth.

As the demand for interactive applications is increasing at a sharp rate, the advent of the 5G network has opened up scopes for a plethora of such applications with its high-speed internet network ensuring real-time data transmission (such as cloud gaming, VR/AR, and connected vehicles). By merging all these technologies, the functioning bandwidth of the Metaverse can be massively boosted, which will be paramount for the high-definition media transfer responsible for most of the data traffic along with the transmission of metadata produced by the employed IoT devices [28]. In a distributed system, the Metaverse will account for the largest portion of the bandwidth accessible, and it should also be taken into account that there will be competition among the other systems or applications for the bandwidth available. Thus, it is very much expected the capacity of 5G cannot fully support the vast requirement of the Metaverse.

Latency is a major issue in vastly interactive applications, such as cloud and online gaming. There are various factors responsible for the motion-to-photon latency, such as the hardware sensor's recording and the processing time. In systems demanding latency in the millimeter scale, operating system's context-switch frequency is kept between 100 Hz and 1500 Hz. Aspects like memory allotment and copying data between different equipment (such as copying between the memory spaces of the CPU and GPU) also have an impact on the motion-to-photon latency [29]. In such complex applications, the requirement of network connectivity between remote users also introduces further latency. However, 5G networking is expected to reduce latency issues; recent studies reflect that the delay in response time due to the 4G networks is very much similar to that of the radio access network (RAN). For the current being, most of the enhancements are leaping forward from the interaction between the operator core network and gNodeB (gNB) [30]. Thus, it is inevitable to contrive the 5G network connectivity in non-standalone mode, where 5G is utilized by the RAN extending to the gNB; meanwhile, the operator core network operates in 4G. Alongside these, making RAN latency standardized to 4 ms for enhanced mobile broadband and fixing it to 0.5 ms in case of ultrareliable low-latency communication (which is still not executed), the interaction among the core network and gNB responsible for the majority of the round-trip latency (ranging from 10 to 20 ms), with minimal support from the ISP [30]. Similarly, without the direct integration of the 5G gNB with the servers, the benefits obtained from edge computing in comparison to cloud computing get highly disrupted [31, 32]. This is most observable in regions with vast cloud employment. Another way to reduce network latency is to manage the whole end-to-end channel [33] by the content providers and to stretch out to the ISP using network visualization [34–37]. Such an initiative requires a coalition between the ISPs and the content providers. To fully utilize the concept of Metaverse, it is very much required to have a full-fledged coalition between all the participants (ISPs, application programmers, content creators) to construct a firm, low delay, highly efficient network connection.

1.11 THE POTENTIAL APPLICATION OF THE METAVERSE IN REVOLUTIONIZING THE HEALTHCARE SYSTEM

Revolutionizing the healthcare system is one of the key targets in most of the tech advancements, and likewise in the development of the Metaverse, we also look forward to such revolutions. Metaverse which we propose to be an intelligent consolidation of all the sustaining state-of-art technologies can fully change the way healthcare

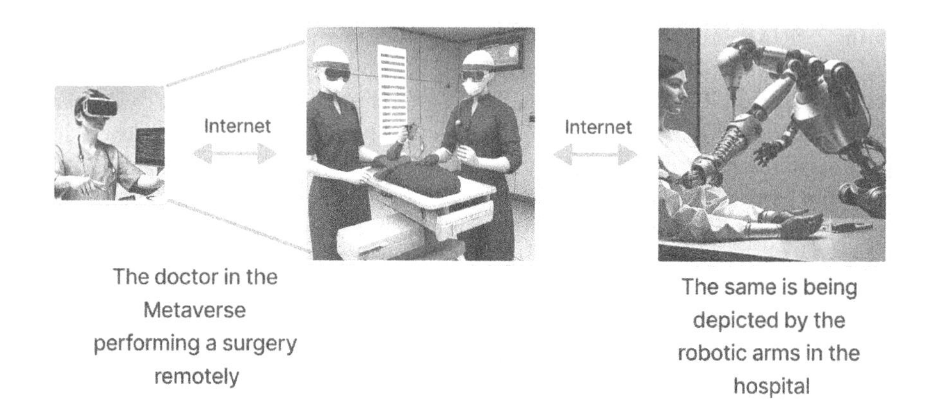

FIGURE 1.15 Application of the Metaverse in the healthcare domain.

systems function. In this section, we propose a simplified high-level model through which the Metaverse can be functional in the medical domain (Figure 1.15).

As we already live in an era where telemedicine has already been implemented, now the next possible upgradation would be introducing remote surgeries, through which surgeons would be enabled to perform surgeries by not being at the site where the patient is. The main advantage of this is that it would enable highly experienced surgeons to collaborate and communicate with each other while performing complicated surgeries, which would keep a major impact in reducing mortality rate and also contribute to reducing surgery costs. Other than this, such techniques can be first adopted by the medical interns for being precise and confident before being exposed to real-world surgeries.

Based on our Metaverse model, the first task would be to create an Avatar that will stimulate the symptoms and potential presence of any comorbidities in the patient placed in the real world, and this is the most challenging task where the application of the AI algorithm is vital, that is, to fully preprocess and analyze all the raw data available to it to take intelligent decisions for articulating the same in the Avatar. The raw data will involve but not be limited to various medical scans, for example, CT scans, ultrasonography in combination with various pathological test reports, leveraging them to deploy a highly efficient AI model for implementation in the Metaverse. The main challenge in integrating such a concept is to fully represent the dynamic nature of the human body in the Avatar, and one of the potential solutions is to closely monitor the patient's internal and external conditions at a high sampling rate via various IoT devices and feeding them into the AI model to take a dynamic decision and reflecting them into the Avatar. Alongside this, the dynamic representation of the Avatar to the doctor working remotely is also immensely important, and this is where the presence of a highly efficient high-speed internet with very low latency is vital because a small network glitch or delay might break the synchronization of the whole process, which would be fatal for the patient. The surgeon at the remote end will be equipped with a VR headset and other AR equipment operating on the Avatar, and the gesture of the surgeon would be synchronized with the robotic arms positioned precisely to operate on the patient in the real world.

1.12 FUTURE RESEARCH ACTIVITIES AND CHALLENGES TO BE HANDLED

1.12.1 Extended Reality

To turn the concept of Metaverse into reality, the implementation of VR/AR/MR is inevitable as an intermediary stage. The VR integration can be regarded as a technical base for the Metaverse creation. As stated earlier, the Metaverse is an expansive virtual ecosystem, where users portrayed as Avatars carry out activities in coordination with real-world entities. The users in the virtual workspace present as digitalized images, similar to that of an ecosystem running concurrently with the real world. A transition to VR will allow the virtual ecosystem to be more immersive and closer to the real world. Alongside these, AR/MR will modernize the physical world even further. Thus, it is very much expected that the future Metaverse will be closely correlated to the real world.

Further technical and design-centric improvements should be focused on shifting the digitalized entities from the VR to the MR environments, and this will lead to the full integration of the virtual entities running in parallel with the real world. Such a technological boost will lead to the creation of super-realistic digitalized entities making their presence known everywhere and interacting with real-world entities through mobile headsets or other user-interactivity techniques.

1.12.2 User Interactivity

The advanced (VR/AR/MR) environments with complex functionality and vast potential cannot replicate all the real-world senses. In such a scenario, the brain–computer interface (BCI) technology could be an amazing standout. BCI functions by creating a direct communication pathway between electronic gadgets and the human brain, therefore bypassing the usage of limbs and language to communicate. As we all know, sensation takes place after the sensory signals pass through the brain, and if the BCI technology is implemented, it will be possible to stimulate various sensory experiences by stimulating specific sensory sections of the brain.

1.12.3 Internet Of Things

IoT devices make use of the XR systems to represent their functionality and allow users to participate in the data management and decision-making process. Representing these functionalities makes it easier for the users to operate with the IoT devices and take specific decisions as per their requirements, which will increase transparency between the users and the employed IoT devices. Alongside this, specific upgradation in the XR interfaces will act as the base to enable user-in-the-loop decision-making. But as of now, designing user interfaces with implementations of dark patterns of IoT, robotics, and precise handling of the new robotic system are still in the primary stage. That is why more research should be promoted to aid the interaction between the Metaverse and IoT devices.

1.12.4 Artificial Intelligence

To give Metaverse the cognizance to operate independently and make the decision with exposure to various environments will promote the concept of self-reliability in the Metaverse, and to achieve such automation to boost user experience and randomness, the integration of AI or more specifically deep learning is inevitable in the Metaverse. The current deep learning models are very resource-demanding and thereby require huge processing strength, which is not at all suitable for implementation in mobile gadgets. Therefore, more research should be put forward to construct efficient models that will be versatile and will require less computational strength to accomplish a plethora of tasks in the Metaverse.

1.12.5 Blockchain

The approach which requires the use of blockchain, the proof of work mechanism, is a pretty slow one and must be improved time complexity-wise to have a practical implementation and achieve more widespread use and be an industry standard. Hence, a faster proof of work or a new algorithm altogether is an industry requirement. As the Metaverse relies on blockchain technology to ensure the authenticity of digital assets and transactions, there is a risk that personal information and data could be compromised. The transparency of the blockchain may be a double-edged sword, as while it ensures transparency and security, it also raises concerns about data privacy and confidentiality.

1.12.6 Computer Vision

CV integration in the Metaverse gives functioning computers the cognizance to interpret/understand the user's spatial configurations with variance to their activities. In order to construct a more precise and genuine virtual environment, the prevailing CV algorithms need to excavate pathways to solve the following issues, that is, the need to interpret sophisticated working environments, connecting the real world with the digitalized objects. Thereby, it is very much expected that the computationally functional structural and scenario interpretive algorithms are to be implemented in the future Metaverse.

1.12.7 Edge Computing

The implementation of MEC requires several third-party entities and service providers. In such a case, there is a huge possibility that the parties would get unauthorized access to the MEC data and exploit it, which would be a huge risk for the users. Therefore, it is very much essential to strengthen security protocol at every layer of the edge ecosystem because a small ignorance of data security would lead to the breakdown of the whole edge ecosystem and thereby the Metaverse services, which will hugely reduce users' trust toward the Metaverse.

1.12.8 Network

One of the main challenges facing Metaverse development is creating a network that can provide a seamless user experience. This requires addressing issues such as latency, throughput, and jitter, all of which are important indicators of network performance. However, in virtual worlds, the complexity of user mobility and concrete perception further complicates this task.

1.13 CONCLUSION

The tech giant Facebook has changed its company name from Facebook to Meta to spread the idea of the Metaverse to a wider community and to lay down an enthusiastic working environment within the company regarding its potential. Other than this, other tech giants like Alphabet and Apple are plotting their roadmaps and are very much interested to implement them soon. With the kind of technological advancements that are already taking place and with the ones that are evolving, it is very much possible that the virtual environment we are now experiencing would be very much ahead in the near future. Therefore, we can expect the technologically fueled world awaiting ahead would be even more connected and vivid due to the presence of powerful computer systems and advanced wearables. However, a lot of technological issues still need to be resolved with great care like networking speed, computational strength, and so on to implement Metaverse in the real world and also to integrate it with our daily life activities.

REFERENCES

1. J. Joshua, "Information bodies: Computational anxiety in Neal Stephenson's snow crash," Interdisciplinary Literary Studies, vol. 19, no. 1, pp. 17–47, 2017.
2. P. Milgram, H. Takemura, A. Utsumi and F. Kishino, "Augmented reality: A class of displays on the reality–virtuality continuum," in Telemanipulator and Telepresence Technologies, vol. 2351. SPIE, 1995, pp. 282–292.
3. A. Bush, "Into the void: Where crypto meets the metaverse," 2021.
4. Y. Jiang, C. Zhang, H. Fu, A. Cannavò, F. Lamberti, H. Y. Lau and W. Wang, "Handpainter-3D sketching in VR with hand-based physical proxy," in Proceedings of the 2021 CHI Conference on Human Factors in Computing Systems, 2021, pp. 1–13.
5. S. Singhal and M. Zyda, Networked Virtual Environments: Design and Implementation. ACM Press/Addison-Wesley Publishing Co.: Boston, MA, 1999.
6. H. Liu, M. Bowman and F. Chang, "Survey of state melding in virtual worlds," ACM Computing Surveys (CSUR), vol. 44, no. 4, pp. 1–25, 2012.
7. R. T. Azuma, "A survey of augmented reality," Presence: Teleoperators & Virtual Environments, vol. 6, no. 4, pp. 355–385, 1997.
8. C. H. Godoy Jr., "Augmented reality for education: A review," *arXiv Preprint*. arXiv:2109.02386, 2021.
9. L.-H. Lee and P. Hui, "Interaction methods for smart glasses: A survey," IEEE Access, vol. 6, pp. 28712–28732, 2018.

10. C. Zhang, A. Bedri, G. Reyes, B. Bercik, O. T. Inan, T. E. Starner and G. D. Abowd, "Tapskin: Recognizing on-skin input for smartwatches," in Proceedings of the 2016 ACM International Conference on Interactive Surfaces and Spaces, 2016, pp. 13–22.
11. Z. Xu, P. C. Wong, J. Gong, T.-Y. Wu, A. S. Nittala, X. Bi, J. Steimle, H. Fu, K. Zhu and X.-D. Yang, "Tiptext: Eyes-free text entry on a fingertip keyboard," in Proceedings of the 32nd Annual ACM Symposium on User Interface Software and Technology, 2019, pp. 883–899.
12. I. Poupyrev, N.-W. Gong, S. Fukuhara, M. E. Karagozler, C. Schwesig and K. E. Robinson, "Project jacquard: Interactive digital textiles at scale," in Proceedings of the 2016 CHI Conference on Human Factors in Computing Systems, 2016, pp. 4216–4227.
13. V. Becker, F. Rauchenstein and G. Sörös, "Connecting and controlling appliances through wearable augmented reality," Augmented Human Research, vol. 5, pp. 1–16, 2020.
14. C. Bermejo Fernandez, L. H. Lee, P. Nurmi and P. Hui, "Para: Privacy management and control in emerging IoT ecosystems using augmented reality," in Proceedings of the 2021 International Conference on Multimodal Interaction, 2021, pp. 478–486.
15. G. Alce, M. Roszko, H. Edlund, S. Olsson, J. Svedberg and M. Wallergard, "[Poster] AR as a user interface for the internet of things—Comparing three interaction models," in 2017 IEEE International Symposium on Mixed and Augmented Reality (ISMAR-Adjunct). IEEE, 2017, pp. 81–86.
16. Y. Cao, Z. Xu, F. Li, W. Zhong, K. Huo and K. Ramani, "V.Ra: An in-situ visual authoring system for robot-IoT task planning with augmented reality," in Proceedings of the 2019 on Designing Interactive Systems Conference, 2019, pp. 1059–1070.
17. L. Chen, A. Ebi, K. Takashima, K. Fujita and Y. Kitamura, "Pinpointfly: An egocentric position-pointing drone interface using mobile AR," in SIGGRAPH Asia 2019 Emerging Technologies, 2019, pp. 34–35.
18. C. Cadena, L. Carlone, H. Carrillo, Y. Latif, D. Scaramuzza, J. Neira, I. Reid and J. J. Leonard, "Past, present, and future of simultaneous localization and mapping: Toward the robust-perception age," IEEE Transactions on Robotics, vol. 32, no. 6, pp. 1309–1332, 2016.
19. R. Mur-Artal and J. D. Tardo´s, "ORB-SLAM2: An open-source slam system for monocular, stereo, and RGB-D cameras," IEEE Transactions on Robotics, vol. 33, no. 5, pp. 1255–1262, 2017.
20. D. G. Lowe, "Distinctive image features from scale-invariant keypoints," International Journal of Computer Vision, vol. 60, pp. 91–110, 2004.
21. E. Rublee, V. Rabaud, K. Konolige and G. Bradski, "Orb: An efficient alternative to sift or surf," in 2011 International Conference on Computer Vision. IEEE, 2011, pp. 2564–2571.
22. S. Milz, G. Arbeiter, C. Witt, B. Abdallah and S. Yogamani, "Visual slam for automated driving: Exploring the applications of deep learning," in Proceedings of the IEEE Conference on Computer Vision and Pattern Recognition Workshops, 2018, pp. 247–257.
23. G. Reitmayr, T. Langlotz, D. Wagner, A. Mulloni, G. Schall, D. Schmalstieg and Q. Pan, "Simultaneous localization and mapping for augmented reality," in 2010 International Symposium on Ubiquitous Virtual Reality. IEEE, 2010, pp. 5–8.
24. T. R. Gadekallu, T. Huynh-The, W. Wang, G. Yenduri, P. Ranaweera, Q.-V. Pham, D. B. da Costa, and M. Liyanage, "Blockchain for the metaverse: A review," *arXiv Preprint.* arXiv:2203.09738, 2022.
25. S. Nakamoto, "Bitcoin: A peer-to-peer electronic cash system," Decentralized Business Review, p. 21260, 2008.
26. B. H. Allah and I. Abdellah, "MEC towards 5G: A survey of concepts, use cases, location tradeoffs," Transactions on Machine Learning and Artificial Intelligence, vol. 5, no. 4, 2017.

27. N. Mohan, A. Zavodovski, P. Zhou and J. Kangasharju, "Anveshak: Placing edge servers in the wild," in Proceedings of the 2018 Workshop on Mobile Edge Communications, 2018, pp. 7–12.
28. S. Lu, Y. Yao and W. Shi, "Collaborative learning on the edges: A case study on connected vehicles," in USENIX Workshop on Hot Topics in Edge Computing (HotEdge), 2019.
29. J. Hestness, S. W. Keckler and D. A. Wood, "GPU computing pipeline inefficiencies and optimization opportunities in heterogeneous CPU–GPU processors," in 2015 IEEE International Symposium on Workload Characterization. IEEE, 2015, pp. 87–97.
30. X. Yuan, M. Wu, Z. Wang, Y. Zhu, M. Ma, J. Guo, Z.-L. Zhang and W. Zhu, "Understanding 5G performance for real-world services: A content provider's perspective," in Proceedings of the ACM SIGCOMM 2022 Conference, 2022, pp. 101–113.
31. L. Corneo, M. Eder, N. Mohan, A. Zavodovski, S. Bayhan, W. Wong, P. Gunningberg, J. Kangasharju and J. Ott, "Surrounded by the clouds: A comprehensive cloud reachability study," in Proceedings of the Web Conference 2021, 2021, pp. 295–304.
32. L. Corneo, N. Mohan, A. Zavodovski, W. Wong, C. Rohner, P. Gunningberg and J. Kangasharju, "(How much) can edge computing change network latency?" in 2021 IFIP Networking Conference (IFIP Networking). IEEE, 2021, pp. 1–9.
33. M. Ammar, E. Zegura and Y. Zhao, "A vision for zero-hop networking (ZEN)," in 2017 IEEE 37th International Conference on Distributed Computing Systems (ICDCS). IEEE, 2017, pp. 1765–1770.
34. K. Diab and M. Hefeeda, "Joint content distribution and traffic engineering of adaptive videos in TELCO-CDNS," in IEEE INFOCOM 2019-IEEE Conference on Computer Communications. IEEE, 2019, pp. 1342–1350.
35. A. Bandyopadhyay, A. Sarkar, S. Swain, D. Banik, A. E. Hassanien, S. Mallik, A. Li and H. Qin, "A game-theoretic approach for rendering immersive experiences in the metaverse", Mathematics, vol. 11, no. 6, p. 1286, 2023.
36. A. Sihna, H. Raj, R. Das, A. Bandyopadhyay, S. Swain and S. Chakrborty, "Medical education system based on metaverse platform: A game theoretic approach", In IEEE 4th International Conference on Intelligent Engineering and Management (ICIEM 2023), 2023, pp. 1–6.
37. P. Gupta, K. Bhadani, A. Bandyopadhyay, D. Banik and S. Swain, "Impact of Metaverse in the near 'future'", In IEEE 4th International Conference on Intelligent Engineering and Management (ICIEM 2023), 2023, pp. 1–6.

2 Digital Transformation in Healthcare

An Engineering Perspective on Metaverse, Game Theory, Artificial Intelligence, and Blockchain

B. Sathyasri
Department of Electronics and Communication Engineering, Vel Tech Rangarajan Dr. Sagunthala R&D Institute of Science and Technology, Chennai, India

G. Aloy Anuja Mary
Department of Electronics and Communication Engineering, Vel Tech Rangarajan Dr. Sagunthala R&D Institute of Science and Technology, Chennai, India

K. Aanandha Saravanan
Department of Electronics and Communication Engineering, Vel Tech Rangarajan Dr. Sagunthala R&D Institute of Science and Technology, Chennai, India

Murali Kalipindi
Department of Artificial intelligence and Machine learning, Vijaya Institute of Technology for Women, Vijayawada, India

2.1 INTRODUCTION

Digital transformation in healthcare is a complex, multifaceted phenomenon. It involves the application and integration of numerous digital technologies, each bringing their unique potential and challenges. These technologies – notably artificial intelligence (AI), the Internet of Things (IoT), blockchain, and the nascent metaverse – are radically altering the traditional healthcare paradigm [1, 2]. As these digital innovations permeate the healthcare sector, they bring a paradigm shift in

DOI: 10.1201/9781003449256-2

service delivery, transforming the landscape in unprecedented ways [3]. The evolution of healthcare services, propelled by these digital technologies, is far-reaching and transformative [4]. These advances offer the promise of personalizing healthcare delivery, tailoring interventions to individual patient characteristics [5]. They bring enhanced levels of precision, predictive capabilities, and efficiency to patient care and health systems management [6]. AI algorithms, for instance, aid in diagnosis, disease progression modeling, and personalized treatment planning, while IoT devices facilitate continuous patient monitoring and data collection, even in remote settings.

Blockchain technology addresses security concerns, particularly in managing and sharing patient data. By enabling transparent, immutable, and decentralized data recording, blockchain presents a solution to the longstanding problem of data breaches in healthcare [7]. Furthermore, the concept of the metaverse – a collective virtual shared space, created by the convergence of physically virtually enhanced reality, holds significant implications for healthcare [8, 9]. From providing immersive training for healthcare professionals to enabling virtual patient consultations, the metaverse presents vast unexplored potentials in healthcare [10]. The COVID-19 pandemic highlighted the urgency and relevance of digital transformation in healthcare. As healthcare institutions grappled with unprecedented demand and the need for infection control, digital technologies provided essential solutions [11, 12]. Telemedicine, powered by AI and IoT, ensured the continuity of care while abiding by social distancing measures. Furthermore, digital health applications supported public health measures such as contact tracing, quarantine management, and disseminating reliable information [13].

The discipline of engineering, once predominantly associated with the creation and optimization of physical systems, has expanded its horizons significantly [14, 15]. Today, it finds transformative applications within the healthcare sector, most notably because of the advent and proliferation of medical devices, robotics, and biomechanical systems [16]. Engineers are no longer confined to traditional roles; they are now instrumental in the creation, optimization, and adaptation of various healthcare technologies. These include but are not limited to robotic surgical systems, wearable health monitors, prosthetics, and even implantable devices that require meticulous design and precise functionality [17].

Engineers bring a unique skill set to the healthcare setting. Their extensive knowledge in systems design, material science, dynamics, and thermodynamics allows them to grasp the nuances of integrating digital technologies into the healthcare system [18]. They can tackle various tasks ranging from optimizing the structural design of sensors and devices in AI-based diagnostic tools for efficient data collection to creating reliable, safe, and efficient IoT systems that respect patient privacy and safety. The role of engineers also extends to the metaverse, where their expertise can ensure the durability, functionality, and maintenance of medical devices, leading to seamless healthcare service delivery [19, 20]. But to truly appreciate the role of engineering in healthcare, we need to delve deeper into the specifics. A crucial aspect lies in the design and development of medical devices. The growing reliance on AI and IoT technologies in the healthcare sector necessitates the creation of devices that can capture, process, and transmit a vast array of data accurately and securely. Here, engineers play an indispensable role by designing ergonomic, efficient, and durable devices that integrate complex digital technologies into usable, patient-friendly formats.

Engineers also bring their expertise to the realm of surgical robotics. These advanced systems can enhance precision, reduce invasiveness, and thus improve patient outcomes. However, designing such systems requires a nuanced understanding of human anatomy, surgical procedures, and principles of robotics – a task engineers are well-suited for. They ensure that these robotic systems are user-friendly, safe, and effective in the surgical environment [21, 22]. Furthermore, the realm of biomechanics and prosthetics is heavily reliant on engineering principles. From designing prosthetic limbs that mirror the functionality of biological counterparts to creating biomechanical models that can predict and analyze human movement, the role of engineering is undeniable [23]. Engineers ensure that these devices and models are not just scientifically sound but also practical and comfortable for the users.

The expanding metaverse, a collective virtual shared space, is also becoming an arena for engineers [24, 25]. Whether it's about designing virtual reality (VR) systems for medical training or creating virtual healthcare facilities for remote patient care, engineers ensure that these systems are robust, reliable, and capable of delivering high-quality virtual experiences [26, 27]. In essence, the role of engineering in healthcare goes beyond mere device design or system optimization. It encompasses understanding the needs of patients, healthcare professionals, and the overall health system and using this understanding to shape digital technology integration. It is about ensuring that technology serves its purpose of enhancing health outcomes and improving the quality of healthcare delivery [28, 29].

This chapter explores the role of engineering in the digital transformation of healthcare, with a focus on metaverse, game theory, AI, IoT, and blockchain technologies. It provides a comprehensive understanding of these technologies and discusses how engineering principles can be applied to enhance their efficacy and integration into healthcare. It also highlights the practical implications of these technologies through real-world case studies and discusses the challenges and future directions of this interdisciplinary field. Figure 2.1 shows the flowchart representing the structural overview of the chapter.

2.2 UNDERSTANDING THE TECHNOLOGICAL LANDSCAPES

2.2.1 Introduction to Metaverse

Originally a concept emerging from the realm of science fiction, penned by the esteemed author Tally Jr [30], the Metaverse now stands as a pioneering idea within the digital landscape. It constitutes a symbiotic confluence of virtual and physical reality, giving rise to an all-encompassing, shared virtual space. The operation of this transformative arena is facilitated by networked computers, which work in concert to create an environment that transcends the limitations of geographical and physical confines [31–34]. Figure 2.2 illustrates the various applications and implications of the metaverse in the healthcare sector. The Metaverse provides a platform for individuals to engage with a dynamically generated environment and interact with

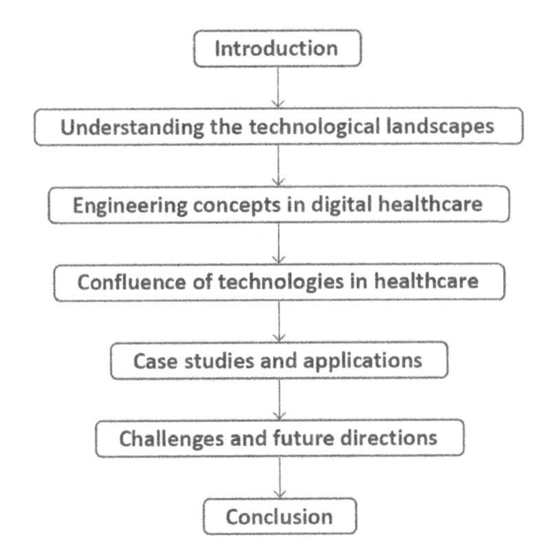

FIGURE 2.1 Flowchart representing the structural overview of the chapter.

other users in real-time. It incorporates a multitude of virtual elements, all of which are designed to replicate the depth and interactivity of the physical world, thereby providing users with a deeply immersive experience. Its potential to instigate change and revolutionize a multitude of sectors cannot be overstated. One such sector that stands to benefit significantly from the rise of the Metaverse is healthcare.

Indeed, the Metaverse has the capacity to revolutionize healthcare as we know it. It paves the way for interactive patient care experiences that go above and beyond what traditional healthcare delivery can provide. Through the Metaverse, medical professionals can engage with patients in an environment that is as close to the physical world as possible, without being constrained by the barriers of distance or time. This becomes especially valuable in the age of telemedicine, where consultations are becoming increasingly digitized and remote [35]. Moreover, the Metaverse extends its benefits to the field of medical education and training. By facilitating virtual simulations, it provides an immersive, risk-free platform for medical practitioners to practice and hone their skills. These simulations can emulate a broad spectrum of scenarios, from common surgical procedures to complex, high-stakes operations. This not only enhances the learning experience but also prepares healthcare professionals for real-life scenarios, thereby improving patient outcomes and overall healthcare quality [36].

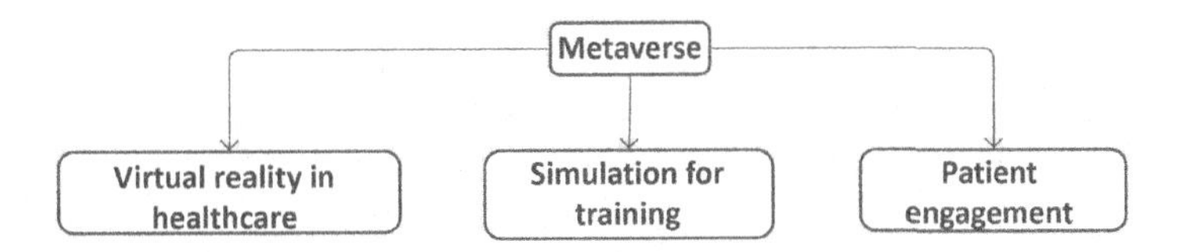

FIGURE 2.2 Diagram illustrating the various applications and implications of the metaverse in the healthcare sector.

But the benefits of the Metaverse in healthcare don't stop there. As it continues to evolve, it is expected to introduce new possibilities and functionalities. For instance, it could allow for real-time monitoring of patients' health status in the virtual environment, thereby enabling healthcare professionals to respond promptly to health changes. Additionally, it could facilitate virtual support groups where patients could interact with peers experiencing similar health conditions, fostering a supportive community and improving mental well-being. In essence, the Metaverse has the potential to redefine the contours of healthcare, providing innovative solutions that enhance patient care, improve healthcare outcomes, and contribute to the overall efficiency of the healthcare system. As we delve deeper into this review, we will explore the role of engineering in harnessing these potentials, thereby unlocking new possibilities for digital healthcare.

2.2.2 Unpacking Game Theory

Game theory, a mathematical discipline of vast potential and application, offers a sophisticated framework for analyzing strategic scenarios. It focuses on competitive situations wherein the outcome of one participant's choice of actions is inextricably linked to the decisions of other participants. Its origins lay within the field of mathematics but its applications have permeated far beyond, spilling into the domains of economics, political science, and more recently, healthcare, among others [37]. The unique value proposition of game theory lies in its capacity to model and predict outcomes in situations characterized by interdependent decision-making. In essence, it provides a roadmap of possible scenarios, dictated by the strategic choices of all involved parties. This predictive capacity of game theory has lent it utility in a wide array of disciplines, but perhaps one of the most promising applications lies in the realm of healthcare.

In healthcare, game theory has emerged as an indispensable tool for driving operational efficiency and improving patient outcomes. It provides a robust framework for strategizing and optimizing various aspects of healthcare delivery. For instance, it can be applied to design optimal strategies for resource allocation – a critical aspect of healthcare management. With limited resources and an ever-growing demand for healthcare services, the need for efficient allocation strategies is paramount. Game theory can provide insights into how these resources can be most effectively distributed to maximize patient care and minimize waste [38]. Patient scheduling, another critical aspect of healthcare operation, also stands to benefit from the application of game theory. Traditional scheduling approaches often struggle with balancing patient needs and operational efficiency. Game theory, with its focus on strategic decision-making, can help design scheduling strategies that take into account both the needs of the patients and the operational constraints of the healthcare facility, leading to improved patient satisfaction and efficient use of resources.

Game theory is not only limited to operational aspects of healthcare; it also extends to the realm of clinical decision-making. By modeling the choices of patients, doctors, and other healthcare professionals as strategic decisions, game theory can help predict health behaviors and outcomes. This can be particularly useful in understanding patient adherence to treatment plans, doctor–patient interactions, and health policy decisions. Moreover, the use of game theory in healthcare is not just limited to these areas. It holds potential in many other aspects, such as pricing strategies for healthcare services, design of health insurance policies, and the development of public health interventions. As the field continues to evolve, it is anticipated that the application of game theory in healthcare will become increasingly sophisticated, driving innovations that enhance healthcare delivery and patient outcomes. In this review, we further delve into the role of game theory in the integration of digital technologies in healthcare from an engineering perspective. We explore how the principles of game theory can guide the design and operation of AI, IoT, and blockchain systems in healthcare, opening up new possibilities for efficient, personalized, and secure healthcare delivery.

2.2.3 Artificial Intelligence in Healthcare

AI in healthcare signifies the strategic employment of complex algorithms and software to emulate human cognition in the comprehension, analysis, and interpretation of multifaceted healthcare and medical data. This computational technology has begun to permeate virtually every facet of the healthcare industry, transforming the way care is delivered, diseases are diagnosed, and treatments are developed. Its potential in enhancing healthcare services is vast and multifaceted [39]. Figure 2.3 shows the visualization of the diverse roles of AI in healthcare. The primary intent of health-related AI applications is to discern and analyze relationships between preventative or therapeutic strategies and their impact on patient outcomes. This insight can enable healthcare providers to predict how a disease might progress, how different individuals might respond to the same treatment, or how a particular patient might benefit from personalized therapeutic regimens, thereby facilitating the shift from reactive to proactive healthcare [40]. Moreover, AI can help drive precision medicine initiatives by using sophisticated algorithms to analyze genomic and clinical data, potentially revealing novel patterns and associations that may not be apparent to the human eye. By stratifying patients based on their unique genetic and clinical profiles, AI can aid in the development of more effective and personalized treatment strategies [41]. AI also plays an increasingly pivotal role in disease diagnosis. Machine learning algorithms can be trained to analyze medical images and electronic health records, enabling them to identify patterns indicative of disease, often with comparable or even superior accuracy to human experts. Such AI-driven diagnostic tools can enhance the efficiency and reliability of disease detection, particularly in resource-limited settings [42].

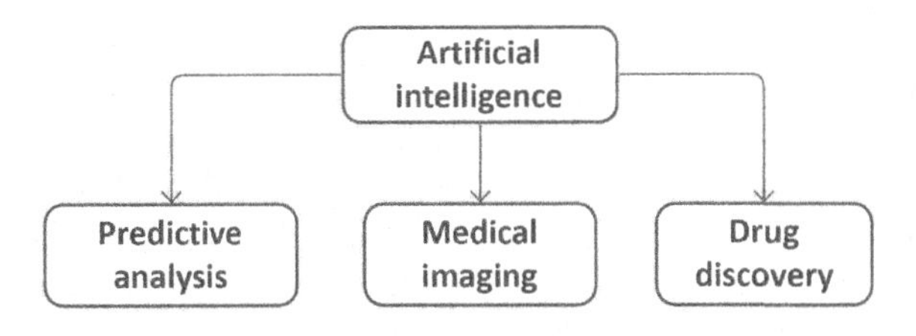

FIGURE 2.3 Visualization of the diverse roles of artificial intelligence in healthcare.

In the context of drug discovery and development, AI can help streamline the process, reduce costs, and increase the success rate of drug candidates [43]. By leveraging AI to analyze vast amounts of biological, chemical, and clinical data, researchers can predict how different compounds might interact with various biological targets, thereby identifying promising drug candidates more efficiently [44]. AI is not just transforming the clinical aspects of healthcare but is also reshaping the operational and administrative facets of the industry. It has been instrumental in optimizing logistics within healthcare settings, such as scheduling appointments, managing patient flow, and predicting patient demand, thereby enhancing operational efficiency and patient satisfaction. In this chapter, we explore how engineering principles can be applied to enhance the design and operation of AI systems in healthcare, particularly in the context of integrating digital technologies like IoT and blockchain. We delve into how engineers can optimize the structural design of AI-based diagnostic tools, develop more efficient and reliable AI-powered medical devices, and navigate the challenges associated with integrating AI into the healthcare sector.

2.2.4 Role of IoT in Digital Healthcare

The IoT, especially its specialized application in healthcare known as the Internet of Medical Things (IoMT), signifies an interconnected ecosystem of medical devices and applications. These tools, seamlessly linked via online computer networks, are transforming the healthcare landscape by enhancing data accessibility, promoting patient-centric care, and improving the overall efficiency of healthcare services. IoMT devices are usually furnished with advanced sensor technology that captures and transmits a wealth of data in real-time. These data are then analyzed and leveraged to generate valuable insights into patients' health conditions, making remote patient monitoring a reality. With this technology, clinicians can keep track of vital patient statistics from afar, thus enabling continuous patient care even outside the traditional healthcare setting. This capacity is particularly advantageous for managing chronic diseases, reducing hospital readmission rates, and delivering elderly care [45]. However, the application of IoT in healthcare extends beyond patient monitoring. Predictive maintenance of medical equipment is another significant area where IoT is making strides. By monitoring the operational parameters and performance of medical devices, IoT can predict potential equipment failures or maintenance needs, thus reducing equipment downtime and preventing unforeseen disruptions in medical services [46].

Moreover, IoT also contributes significantly to the optimization of patient care in healthcare facilities. For instance, IoT-enabled smart beds can detect when they are occupied and when a patient attempts to get up, thereby helping to prevent falls. Similarly, IoT devices can monitor and manage the environmental conditions within healthcare facilities, such as temperature, humidity, and air quality, which can impact patient comfort and health outcomes [47]. IoT is also proving to be a game-changer in the realm of healthcare logistics. Real-time tracking of medical assets, monitoring of storage conditions for sensitive medical supplies, automated inventory management, and efficient waste management are just a few examples of how IoT can enhance operational efficiency in healthcare settings [48]. Lastly, IoT's

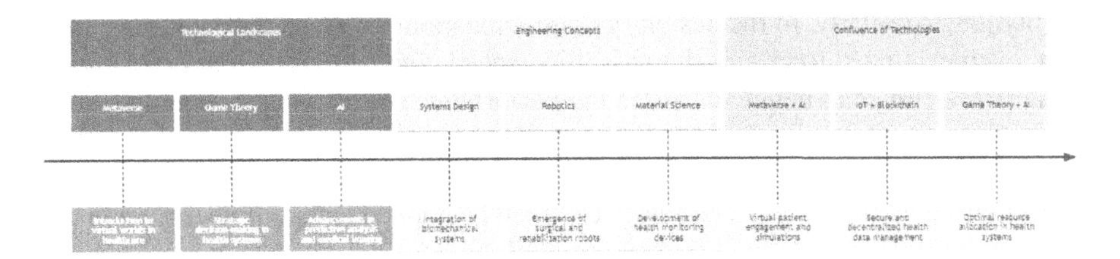

FIGURE 2.4 Timeline showcasing the evolution and integration of various technological and engineering concepts in the realm of digital healthcare.

potential in supporting public health initiatives is worth highlighting. For instance, IoT devices can collect large volumes of health-related data from the population, which can be analyzed to track disease trends, monitor the spread of infectious diseases, and inform public health policies [49]. In this chapter, we will explore how the principles and techniques of engineering can be applied to the design and operation of IoT systems in healthcare. We will delve into the challenges associated with the integration of IoT into healthcare and discuss the potential role of engineers in overcoming these challenges and maximizing the potential of this technology. Figure 2.4 shows a typical timeline of the evolution and integration of various technological and engineering concepts in the realm of digital healthcare.

2.3 ENGINEERING CONCEPTS AND TECHNIQUES IN DIGITAL HEALTHCARE

2.3.1 Systems Design and Analysis

A crucial facet of engineering in digital healthcare revolves around the principles of systems design and analysis. The ability of engineers to conceptualize, design, and execute solutions for complex problems forms the foundation of their significant contribution to the healthcare sector [50, 51]. In this respect, engineers employ a range of techniques such as modeling, simulation, and optimization to engineer effective solutions that address multifaceted health-related problems [52]. The design and analysis of digital healthcare systems, in particular, bring to the forefront the expertise of engineers [53, 54]. Given the complexity and diversity of digital healthcare, it becomes imperative to design systems that can effectively capture, communicate, and process data while ensuring the system's robustness and reliability [55]. Engineers, through their expertise in systems design and analysis, can develop models that accurately represent the system's operation, perform simulations to evaluate system performance under varying conditions, and optimize system parameters to enhance efficiency and reliability [56].

These principles come into play in the design of various elements of digital healthcare systems. For example, in designing a data collection mechanism, engineers need to consider factors such as sensor placement, accuracy, and precision [15, 57]. They may use simulation techniques to test different sensor configurations and optimization algorithms to find the optimal sensor arrangement that maximizes data quality [58]. Similarly, in the design of communication protocols for data transmission, engineers need to ensure the reliability, speed, and security of data exchange. System modeling can help understand the performance of different communication protocols under varying network conditions, while simulation can be used to evaluate the impact of network congestion, data loss, and security breaches on system performance. Optimization techniques can then be applied to fine-tune the communication protocols for optimal performance [59].

Moreover, the analysis of information processing strategies is another area where engineering principles play a crucial role. Given the massive volume of data generated in digital healthcare systems, engineers need to design efficient algorithms that can process these data in real-time. This involves the use of models to understand computational requirements, simulations to evaluate algorithm performance, and optimization to improve processing speed and accuracy [60, 61]. In sum, the principles and techniques of systems' design and analysis in engineering provide the necessary tools and frameworks to address the challenges in designing and implementing robust and efficient digital healthcare systems. This chapter will further delve into these techniques and their applications in different aspects of digital healthcare.

2.3.2 Robotics and Automation

Robotics and automation, a profound subset of engineering, have made a significant impact in the healthcare field, transforming multiple areas ranging from surgical interventions to administrative tasks [62]. The integration of robotic systems into healthcare has undoubtedly changed the way surgical procedures are carried out [63, 64]. Their ability to operate with immense precision and repeatability has brought about a paradigm shift in how surgical interventions are viewed, contributing significantly to enhancing surgical outcomes [65]. Robotic surgery systems, such as the well-known da Vinci Surgical System, offer surgeons enhanced dexterity, greater precision, and better control. This system allows for minimally invasive procedures, a stark contrast to conventional open surgeries, leading to a reduction in surgical site complications, patient discomfort, and recovery time [66].

Yet, the role of robotics is not confined to surgical interventions alone. They have equally demonstrated utility in diagnostics, patient care, and rehabilitation. In diagnostics, robots can conduct repetitive tasks with high precision, reducing the probability of human errors. Robotic devices, such as blood-drawing robots and automated diagnostic machines, ensure more accurate and reliable test results [67]. In patient care, robots can assist healthcare professionals in handling day-to-day tasks, reducing their workload and ensuring better care for patients. For instance, robotic nurses can perform routine tasks such as administering medication, taking vitals, and moving patients, thereby freeing up time for human nurses to focus on more critical care tasks. Moreover, robots can provide companionship to patients, especially those with cognitive impairments, and can monitor their well-being continuously [68].

Rehabilitation is another domain within healthcare that has been revolutionized by the advent of robotics. Rehabilitation robots can help patients regain lost physical capabilities by providing personalized and consistent therapy. They can adapt to the individual needs of patients and monitor their progress over time, offering feedback that can be used to adjust treatment plans. Notably, robotic exoskeletons have shown promising results in assisting patients with mobility issues, enabling them to lead more independent lives [69]. Meanwhile, automation has optimized healthcare operations, streamlining logistical and administrative tasks. Automated systems can handle repetitive and time-consuming tasks such as patient scheduling, billing, and record-keeping, leading to increased efficiency, reduced errors, and improved patient satisfaction. Additionally, automation of hospital logistics, such as inventory management and delivery of medical supplies, has resulted in operational efficiencies and cost savings [70]. The integration of robotics and automation into healthcare has led to substantial improvements in patient care, operational efficiency, and overall patient outcomes. However, it is crucial to further research and develop these technologies while considering the ethical, social, and legal implications of their use in healthcare.

2.3.3 Biomechanical Systems Integration

Biomechanical systems integration represents a unique interface between mechanical systems and biological structures, where engineering concepts are leveraged to design, optimize, and understand the operation of biomedical systems. This domain encapsulates a variety of applications in healthcare, including the design of prosthetics, implants, and bio-integrated sensors. Prosthetic design and development are core area where engineers' expertise comes into play. Creating prosthetics involves a deep understanding of the human body's mechanics and the application of mechanical principles to design a device that can mimic natural movements. Engineers contribute significantly by optimizing the design to ensure the prosthetic device is comfortable, functional, and user-friendly. For instance, engineers may work on improving the alignment system in lower limb prosthetics to increase stability and reduce energy expenditure during walking [69].

Implantable devices are another area where engineering intersects with healthcare. Devices such as pacemakers, cochlear implants, or hip replacements require engineers' expertise in materials, dynamics, and system design. These professionals ensure that the devices are safe, durable, and capable of functioning within the body's challenging environment [71]. For instance, in the design of hip implants, engineers consider factors such as the device's strength, wear resistance, and compatibility with body tissues. They also work on refining the device's design to provide the best possible mobility and comfort for patients [70]. Bio-integrated sensors, another crucial area in the field of biomechanics, provide real-time health monitoring, playing an integral role in advancing personalized healthcare. These sensors can be attached to the body or even implanted to track vital signs, monitor disease progression, and detect anomalies at an early stage. These sensors' design needs to consider the complex interactions between the sensor material and the biological environment to ensure accurate data collection while minimizing the risk of adverse reactions. Engineers, with their deep understanding of materials and system design, are key players in the development and optimization of these bio-integrated sensors [72].

From prosthetics to implants and bio-integrated sensors, biomechanical systems integration is a testament to the significant role engineering plays in healthcare. This field, however, is continuously evolving, driven by technological advances and a deeper understanding of biological systems. Therefore, engineers must keep pace with this rapidly changing landscape, refining their skills and expanding their knowledge to continue making significant contributions to healthcare.

2.3.4 Material Science for Health Monitoring Devices

Material science's unique intersection with engineering forms the backbone of health monitoring devices' development, fundamentally impacting their functionality, reliability, and user comfort. Engineers, armed with their intricate understanding of material properties, are well positioned to design innovative solutions that meet the stringent requirements of biocompatibility, durability, and precision in health monitoring applications. Biocompatibility is a paramount consideration in material selection for health monitoring devices. The materials must not elicit adverse biological responses when in contact with body tissues or fluids, thereby ensuring user safety and comfort. This is particularly critical for devices that are meant to be implanted or come into prolonged contact with the skin. Materials like titanium, certain types of stainless steel, and specific polymers and ceramics are commonly used due to their biocompatible nature and favorable mechanical properties [73]. Durability is another crucial attribute for materials used in health monitoring devices. These devices must withstand mechanical stresses, thermal fluctuations, and exposure to body fluids or sweat, all while maintaining their functionality over extended periods. Engineers' factor in these considerations during the design phase, selecting materials that can resist wear and tear, corrosion, and other forms of degradation. This could range from robust metal alloys for implantable devices to flexible, water-resistant polymers for wearable monitors [74].

Accurate and sensitive measurements are the heart and soul of health monitoring devices. Here, material science again plays a significant role. The choice of materials can profoundly influence the device's ability to detect physiological signals and convert them into readable data. For instance, piezoelectric materials that generate an electric charge in response to mechanical stress are widely used in pressure and motion sensors. In contrast, conductive polymers that change their electrical resistance with strain are popular in flexible strain gauges for wearable devices [75]. Advancements in material science, such as flexible electronics and bio-compatible materials, have dramatically broadened the scope and potential of health monitoring devices. Flexible electronics allow for the development of devices that conform to the body's contours, enhancing user comfort and device performance. At the same time, novel biocompatible materials are continually being developed, offering improved safety profiles and functionality [76]. Thus, material science's critical role in developing health monitoring devices emphasizes the multidisciplinary nature of modern healthcare solutions. As technologies and materials continue to evolve, engineers must stay abreast of these developments, continually integrating new materials and concepts into their designs to develop innovative, effective, and safe health monitoring devices.

2.4 CONFLUENCE OF METAVERSE, AI, IoT, AND BLOCKCHAIN IN HEALTHCARE: AN ENGINEERING PERSPECTIVE

2.4.1 Metaverse and Its Implications for Engineering in Healthcare

The Metaverse, characterized by the convergence of virtual and augmented reality, signals a new frontier in healthcare innovation, significantly influencing engineering's role within this realm. This integrative virtual shared space, continuously built and maintained by networked computers, offers a fertile ground for the engineering domain to bridge the gap between digital healthcare aspirations and their tangible realizations. A promising application of the Metaverse in healthcare is the creation of immersive virtual training environments, which can drastically improve the proficiency of healthcare professionals. These virtual environments, mimicking real-world scenarios, provide an avenue for clinicians to practice complex surgical procedures or diagnostic techniques without the risks associated with actual patients. The responsibility lies with engineers to design these environments with a high degree of realism. This entails developing and integrating haptic feedback systems to simulate the feel of tissues, designing user-friendly interfaces for virtual tools, and optimizing the system's performance to ensure a seamless, lag-free experience. They are also charged with incorporating safety mechanisms to prevent any potential harm that could arise from malfunctions or incorrect usage of these virtual systems [77]. Engineers also play a pivotal role in revolutionizing patient care through the Metaverse. This can be achieved by creating interactive virtual platforms for therapy, wellness, and patient education. Such platforms can facilitate mental health therapy sessions in comforting environments, provide interactive physical therapy exercises, and deliver comprehensive educational modules for patients to understand their conditions better. Engineers contribute by designing comfortable and ergonomic interfaces, enhancing the realism of the virtual environments, and ensuring the robustness and safety of the systems. They also deal with challenges such as reducing motion sickness associated with extended VR sessions, enhancing the mobility of users within the virtual environment, and ensuring the systems' privacy and security [78].

Moreover, engineers can aid in developing virtual replicas of medical devices within the Metaverse, enabling remote consultation, troubleshooting, and even maintenance. They can model the functionality and mechanics of these devices, creating accurate digital twins that can be manipulated and studied in the virtual space. This has significant implications for remote areas where access to advanced healthcare technology and expertise is limited [56]. In essence, the Metaverse represents a paradigm shift in how healthcare services are delivered and consumed. With its promise of immersive, interactive, and customizable healthcare experiences, the Metaverse holds the potential to democratize access to high-quality healthcare and improve patient outcomes. As engineers continue to navigate this exciting landscape, they will undoubtedly play a critical role in actualizing the immense potential of the Metaverse in healthcare. Figure 2.5 shows a typical user's journey in a digital healthcare platform.

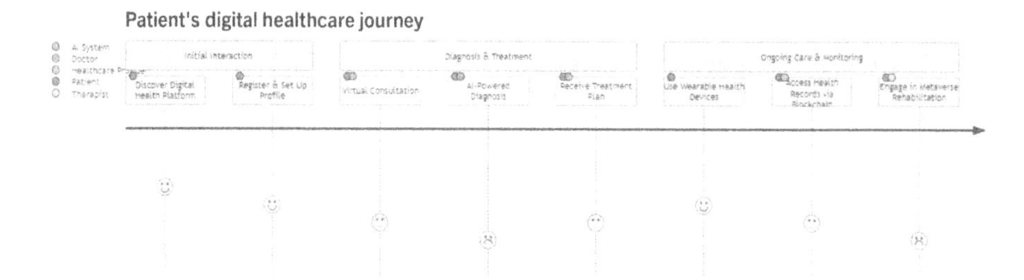

FIGURE 2.5 A user journey diagram illustrating a patient's interaction and experience with a comprehensive digital healthcare platform.

2.4.2 Role of Game Theory in Designing AI and IoT Systems

Game theory, a mathematical framework for analyzing competitive scenarios where the outcome of a participant's choice relies upon the actions of other participants, has profound implications for designing AI and IoT systems within healthcare. Engineers, with their analytical acuity and design expertise, play a pivotal role in employing game theory's strategic and decision-making principles to optimize these systems. The potency of game theory is exhibited when optimizing resource allocation within healthcare IoT networks, a task that mirrors the allocation of scarce resources in competitive scenarios [79]. These networks, comprising a multitude of interconnected nodes like wearable devices, implantable sensors, and medical imaging equipment, are continuously engaged in transmitting health-related data. Each device or node requires resources, such as bandwidth or power, to perform its function, often leading to competition in resource-limited environments.

Game theory provides a framework for making strategic decisions in these competitive scenarios. For instance, how should limited bandwidth be assigned among several devices vying for network access? Or, how should power resources be distributed among multiple nodes to ensure their efficient operation? Engineers can utilize game theoretic models to solve these problems, ensuring the optimal allocation of resources that maximizes network performance and minimizes potential bottlenecks [80]. Similarly, the principles of game theory can influence the design and operation of AI algorithms in healthcare. These algorithms often need to balance multiple factors, such as maximizing predictive accuracy while minimizing computational cost or personalizing medical treatments by adjusting multiple variables in response to patient data. In this scenario, the application of game theory becomes instrumental. It assists in predicting the interaction of multiple variables and formulates strategic responses to optimize outcomes.

Consider the realm of personalized medicine, where AI algorithms aim to tailor medical treatments based on an individual's genetic makeup, lifestyle, and environment. These algorithms need to balance several factors, such as the effectiveness of the treatment, potential side effects, and the patient's personal preferences. By applying game theoretic models, engineers can optimize these algorithms, ensuring that the interaction of these factors is strategically balanced to yield the best possible treatment plan for the patient [81]. Moreover, the game theory's influence extends to the management and operation of AI systems. For instance, multiple AI agents in a healthcare system may need to coordinate their actions to provide efficient care. These agents may be tasked with different roles, such as diagnosing illnesses, scheduling appointments, or managing hospital resources, and may need to cooperate and compete for resources. Game theory provides the principles for designing strategies for these AI agents, promoting cooperation and competition where needed to ensure the efficient functioning of the healthcare system. Thus, game theory is a potent tool that engineers can leverage in designing and optimizing AI and IoT systems within healthcare. It provides the principles to make strategic decisions in complex, competitive scenarios, ensuring the efficient and effective operation of these systems.

2.4.3 Blockchain's Role in Secure Health Data Management: An Engineering Viewpoint

In an era where digital transformation has pervasively penetrated the healthcare landscape, the management of health data has emerged as a matter of significant concern. The advent of new-age technologies like AI, IoT, and Metaverse has led to a substantial increase in the volume and complexity of health data. This deluge of data brings forth an urgent need for secure, efficient, and transparent management systems [82, 83]. Herein, blockchain technology, with its innate properties of security, decentralization, and transparency, appears to be a promising solution. From an engineering perspective, the integration of blockchain technology with healthcare devices and platforms signifies an exciting realm of opportunities. Blockchain technology, a decentralized ledger system, ensures the security, immutability, and transparency of data transactions. In the context of healthcare, it holds the potential to address significant challenges such as secure data transfer, patient privacy, data interoperability, and consent management [84, 85]. Engineers, with their unique blend of design skills, systems thinking, and technical expertise, can play a crucial role in harnessing this potential. For instance, consider an IoT-based health monitoring device, such as a smart wearable, that continuously collects patient health data like heart rate, blood pressure, and activity levels. These data, often sensitive, need to be securely transmitted and stored, maintaining the patient's privacy while also being accessible for healthcare providers for timely and efficient care [86, 87]. Figure 2.6 presents a schematic representation of the multifaceted applications of blockchain in healthcare.

In such a scenario, engineers could design a system where the collected data are encrypted and recorded on a blockchain [88, 89]. The decentralized nature of blockchain ensures that the data are not stored at a single, vulnerable location but

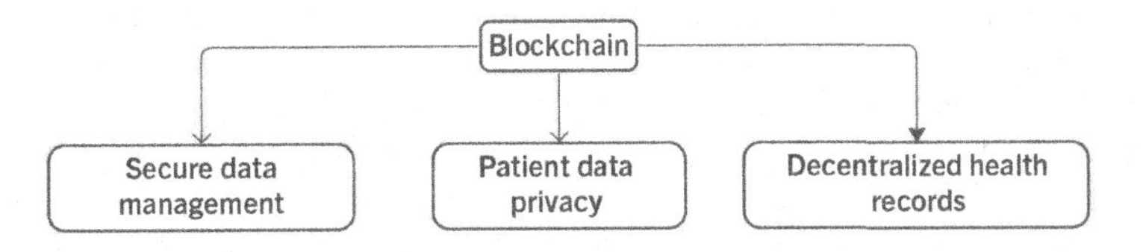

FIGURE 2.6 Schematic representation of the multifaceted applications of blockchain in healthcare.

distributed across multiple nodes, enhancing security. Further, the data recorded on the blockchain are immutable, meaning it cannot be tampered with, thereby ensuring data integrity [90]. Engineers can also utilize their knowledge of systems design and analysis to create robust interfaces between healthcare devices and the blockchain. They can optimize the data transmission protocols, ensuring that the data from the devices are accurately and efficiently recorded on the blockchain. They can also design systems to alert healthcare providers in real-time if any significant deviations in health parameters are detected, thereby facilitating proactive care. Moreover, engineers can address the challenges of data interoperability in healthcare by designing systems that utilize blockchain to link disparate health data sources. This could include data from electronic health records, medical imaging, genomic data, and wearable devices, among others. By providing a unified, secure, and immutable record of a patient's health history, blockchain-based systems could facilitate personalized care, informed decision-making, and improved patient outcomes. The integration of blockchain technology with healthcare devices and platforms is a promising avenue for secure health data management. Engineers, with their broad range of skills and expertise, are uniquely positioned to contribute to this endeavor, designing systems that ensure secure data transfer, patient privacy, and data interoperability [91].

2.5 CASE STUDIES AND REAL-WORLD APPLICATIONS

2.5.1 AI and IoT in Medical Devices

The burgeoning field of digital health has seen a considerable increase in the deployment of AI and IoT technologies in medical devices. These technologies, in tandem, have shown tremendous promise in shaping patient-centered healthcare, facilitating early disease detection, and enhancing the delivery of personalized healthcare solutions. Among the myriad applications of AI and IoT in healthcare, their integration in patient monitoring systems, particularly in smart wearables, glucose monitors, and implantable cardioverter defibrillators, has brought about transformative changes.

Consider the case of the iRhythm Zio XT patch, an exemplar of AI and IoT integration in a wearable device. The Zio XT is a lightweight, easy-to-use wearable patch that records heartbeat data over an extended period. This device integrates AI algorithms that analyze electrocardiogram (ECG) data continuously, detecting any anomalies that could indicate conditions such as arrhythmias. By leveraging AI, the device can sift through large volumes of ECG data, identify relevant patterns, and provide meaningful insights, a task that would be highly time-consuming and prone to error if done manually. Upon detection of any abnormalities, the device, leveraging IoT capabilities, can transmit this information in real-time to clinicians, enabling prompt intervention and patient care. This real-time data transfer, facilitated by IoT, ensures that any critical changes in a patient's heart rhythm are immediately brought to the attention of healthcare providers, reducing the time lag in diagnosis and enhancing the likelihood of successful treatment. In essence, the iRhythm Zio XT patch illustrates the profound implications of AI and IoT integration in healthcare, facilitating real-time health monitoring and the delivery of timely, personalized care [92].

2.5.2 Blockchain in Patient Data Management

As health data becomes increasingly digitized, the secure management and sharing of this information emerge as significant challenges. Blockchain technology, with its decentralized and immutable properties, offers a potential solution to these challenges. A practical illustration of the application of blockchain in patient data management is the MedRec project initiated by the MIT Media Lab. MedRec is a novel, decentralized record management system that leverages the Ethereum blockchain to ensure the secure transfer and storage of patient medical records. The architecture of MedRec is designed such that it assigns patients the authority over their health data, allowing them to grant access to different healthcare providers as needed. This feature promotes a patient-centric approach to healthcare, empowering patients to actively participate in their care process. Furthermore, MedRec ensures interoperability among different healthcare providers. In the current healthcare scenario, a patient's medical records are often scattered across multiple healthcare facilities, making it challenging to have a consolidated view of the patient's health history. By storing patient data on a blockchain, MedRec enables the secure and efficient sharing of patient data among different healthcare providers, providing a comprehensive, unified view of the patient's health history. This feature can facilitate informed clinical decision-making, enable seamless transitions of care, and ultimately lead to improved patient outcomes. The MedRec project, thus, provides a tangible example of how blockchain technology can revolutionize patient data management, fostering a more efficient, secure, and patient-centric healthcare system [93].

2.5.3 Game Theory in Health Systems Design

Game theory, with its ability to model strategic interactions among decision-making entities, provides valuable insights into health systems design. A salient example of this application can be found in the realm of organ transplant allocation, a process that involves complex decision-making and strategic interactions among multiple stakeholders. Organ transplantation is often the last resort for patients with end-stage organ diseases [94, 95]. However, the demand for transplantable organs far exceeds the available supply, necessitating the need for a fair and efficient organ allocation system. Researchers have proposed the use of game theory models to enhance organ allocation policies, striking a balance between efficiency (i.e., maximizing the utilization of available organs) and equity (i.e., ensuring a fair distribution of organs among potential recipients) [96]. Game theory models designed for organ allocation consider hospitals, organ procurement organizations, and patients as players, each with their preferences and strategies. These models aim to understand the strategic behavior of these players under different policy scenarios, thus providing insights into how changes in policies might influence player behavior and overall system performance. By doing so, these models offer a systematic approach to inform policy changes that could lead to optimal organ allocation, improving survival rates and quality of life for recipients. This application of game theory underscores its potential in shaping health policy and influencing health outcomes.

2.5.4 Virtual Reality and Simulation in Metaverse Healthcare

The rise of the Metaverse, characterized by the integration of virtual and augmented reality technologies, holds transformative potential for healthcare. Notably, the use of VR and simulation in the Metaverse has given rise to novel platforms for medical education and training. The Osso VR platform provides a compelling illustration of this application. Osso VR is an innovative surgical training platform that utilizes VR technology to create an immersive, interactive environment for practicing surgical procedures. By donning a VR headset, surgeons can perform complex procedures in a virtual operating room, allowing them to hone their skills without the risks associated with real-world surgical practice.

A significant advantage of the Osso VR platform lies in its capacity to provide immediate feedback. As surgeons navigate through a procedure, the platform evaluates their performance in real-time, pinpointing areas for improvement. This immediate feedback loop facilitates rapid skill acquisition and refinement, ultimately enhancing surgical proficiency. Furthermore, by recreating the high-stakes environment of a real operating room, Osso VR helps surgeons build resilience and adaptability, crucial skills for managing unexpected situations during surgery. Studies have shown that training on platforms like Osso VR can reduce surgical errors and enhance patient safety, underscoring the potential of VR and simulation in the Metaverse for advancing medical education and training [97]. These case studies, encompassing the application of game theory in organ transplant allocation and the role of VR in Metaverse healthcare, offer a glimpse into how emerging technologies and analytical approaches can revolutionize healthcare delivery and outcomes.

2.6 CHALLENGE AND FUTURE DIRECTIONS

2.6.1 Technical Challenges

While the amalgamation of Metaverse, AI, IoT, and blockchain technologies heralds a new era in healthcare, it is not without its share of technical challenges. The very act of integration, the interweaving of these technologies to provide seamless healthcare services, necessitates a robust and secure IT infrastructure. This requirement becomes increasingly critical as the volume and diversity of health-related data surge [98]. IoT devices, the cornerstone of real-time and remote patient monitoring, are particularly vulnerable to security threats. It's essential to build in robust security measures right from the design phase of these devices to prevent data breaches and preserve patient privacy [99]. Blockchain technology, while offering revolutionary potential for secure and decentralized data management, struggles with scalability issues [100, 101]. Its inherently sequential operation makes it challenging to scale up to manage the high transaction volumes typically encountered in healthcare settings. On the AI front, a significant area of concern is the "black box" nature of many AI algorithms. This opaqueness could be problematic, particularly in critical health applications where decision transparency is crucial for trust, validation, and accountability. The development of explainable AI models, which provide understandable reasoning behind their predictions or decisions, is an ongoing research area [102].

2.6.2 Ethical and Regulatory Issues

Besides technical challenges, the use of Metaverse, AI, IoT, and blockchain technologies in healthcare introduces an array of ethical and regulatory issues. As these technologies handle sensitive patient data, concerns surrounding data privacy, confidentiality, and informed consent are heightened. For instance, although blockchain technology provides enhanced security for health data, the challenge lies in ensuring anonymization of the blockchain-stored data while preserving its utility for analysis [103]. Achieving this balance is crucial in maintaining patient privacy and meeting ethical obligations. Moreover, the regulatory landscape for these rapidly evolving technologies is still in flux [104, 105]. Regulators must grapple with the unprecedented challenges these technologies pose, striving to protect patient interests without stifling innovation. This situation calls for a multidisciplinary effort involving technologists, ethicists, legal experts, and policymakers to craft balanced and forward-looking regulations [106].

2.6.3 Prospects and Predictions for Future

Despite these challenges, the future of healthcare, propelled by the integration of Metaverse, AI, IoT, and blockchain technologies, is replete with exciting possibilities. Imagine a future where AI-driven telehealth consultations are commonplace, offering convenient and personalized care to patients irrespective of their geographical location. Consider the potential of IoT-enabled devices to revolutionize remote patient monitoring, enabling real-time health data tracking and timely interventions [107, 108]. Envision a secure healthcare data ecosystem underpinned by blockchain technology, with patients enjoying unprecedented control over their health records. Ponder on the transformative impact of the Metaverse in medical education, with surgeons honing their skills in immersive, VR surgical simulators, or patients benefiting from VR-based therapeutic interventions [109]. Engineers stand at the forefront of this healthcare revolution, tasked with the design and optimization of these technologies and their integration into efficient, user-friendly, and safe health systems. Their contribution will be crucial in shaping the healthcare landscape of the future.

2.7 CONCLUSION

The intersection of Metaverse, AI, IoT, and blockchain has ushered a new paradigm in healthcare, redefining the way care is delivered and accessed. As these technologies continue to evolve, they present novel opportunities for engineers, ranging from the design of complex health systems to the development of advanced medical devices. However, the adoption and integration of these technologies are not without challenges. Technical, ethical, and regulatory concerns pose significant hurdles. Future work should focus on overcoming these challenges to fully leverage the benefits of these technologies in healthcare. Collaboration between engineers, healthcare providers, ethicists, and policymakers will be key to shaping an effective, efficient, and ethically sound digital healthcare future.

REFERENCES

1. R. Thilagavathy, J. Jagadeesan, A. Parkavi, M. Radhika, S. Hemalatha and M. G. Galety, "Digital transformation in healthcare using eagle perching optimizer with deep learning model," Expert Systems: e13390, 2023.
2. F. Teixeira, E. Li, L. Laranjo, C. Collins, G. Irving, M. J. Fernandez, J. Car, M. Ungan, D. Petek, R. Hoffman, A. Majeed, K. Nessler, H. Lingner, G. Jimenez, A. Darzi, C. Jácome and A. L. Neves, "Digital maturity and its determinants in general practice: A cross-sectional study in 20 countries," Frontiers in Public Health, vol. 10, 2023.
3. A. Kapoor, S. Guha, M. Kanti Das, K. C. Goswami and R. Yadav, "Digital healthcare: The only solution for better healthcare during COVID-19 pandemic?" Indian Heart J, vol. 72, pp. 61–64, 2020.
4. H. C. Vollmar, C. Lemmen, U. Kramer, J. G. Richter, M. Fiebig, F. Hoffmann and M. Redaèlli, "Digital transformation of healthcare: A Delphi study of the working groups digital health and validation and linkage of secondary data of the German network for health services research (DNVF) [Digitale transformation des gesundheitswesens—eine delphi-studie der arbeitsgruppen digital health und validierung und linkage von sekundärdaten des deutschen netzwerk versorgungsforschung (DNVF)]," Gesundheitswesen, vol. 84, pp. 581–596, 2022.
5. J. Vidal-Alaball, I. Alarcon Belmonte, R. Panadés Zafra, A. Escalé-Besa, J. Acezat Oliva and C. Saperas Perez, "Approach to digital transformation in healthcare to reduce the digital divide [Abordaje de la transformación digital en salud para reducir la brecha digital]," Atencion Primaria, vol. 55, 2023.
6. S. Berrouiguet, M. M. Perez-Rodriguez, M. Larsen, E. Baca-García, P. Courtet and M. Oquendo, "From eHealth to iHealth: Transition to participatory and personalized medicine in mental health," Journal of Medical Internet Research, vol. 20, p. e2, 2018.
7. R. Miotto, F. Wang, S. Wang, X. Jiang and J. T. Dudley, "Deep learning for healthcare: Review, opportunities and challenges," Brief Bioinformatics, vol. 19, pp. 1236–1246, November 2018.
8. T. Petzold and O. Steidle, "Digital transformation of German healthcare organizations: Current status and existing challenges from the perspective of quality management [Digitale transformation deutscher gesundheitseinrichtungen: Aktueller stand und bestehende herausforderungen aus sicht des qualitätsmanagements]," Bundesgesundheitsblatt—Gesundheitsforschung - Gesundheitsschutz, vol. 66, pp. 972–981, 2023.
9. J. Nitsche, T. S. Busse and J. P. Ehlers, "Teaching digital medicine in a virtual classroom: Impacts on student mindset and competencies," International Journal of Environmental Research and Public Health, vol. 20, 2023.
10. X. Chen, C. He, Y. Chen and Z. Xie, "Internet of Things (IoT)—Blockchain-enabled pharmaceutical supply chain resilience in the post-pandemic era," Frontiers of Engineering Management, vol. 10, pp. 82–95, 2023.
11. F. Santarsiero, G. Schiuma, D. Carlucci and N. Helander, "Digital transformation in healthcare organisations: The role of innovation labs," Technovation, vol. 122, 2023.
12. E. A. Regan, "Changing the research paradigm for digital transformation in healthcare delivery," Frontiers in Digital Health, vol. 4, 2022.
13. N. Lurie and B. G. Carr, "The role of telehealth in the medical response to disasters," JAMA Intern. Med, vol. 178, p. 745, 2018.
14. S. Pal, K. Kalita, A. Majumdar and S. Haldar, "Optimization of frequency separation of laminated shells carrying transversely distributed mass using genetic algorithm," Journal of Vibration and Control, p. 10775463231190277, 2023.

15. K. Kalita, V. Kumar and S. Chakraborty, "A novel MOALO-MODA ensemble approach for multi-objective optimization of machining parameters for metal matrix composites," Multiscale and Multidisciplinary Modeling, Experiments and Design, vol. 6, pp. 179–197, 2023.
16. N. M. Robbins and J. E. Smith, "Conflicts of interest between neurologists and pharmaceutical and medical device industries," in Integrity of Scientific Research: Fraud, Misconduct and Fake News in the Academic, Medical and Social Environment, Springer, 2022, pp. 487–498.
17. R. H. Taylor, A. Menciassi, G. Fichtinger, P. Fiorini and P. Dario, "Medical robotics and computer-integrated surgery," Springer Handbook of Robotics, Springer, 2016, p. 1657–1684.
18. C. Mavroidis, R. G. Ranky, M. L. Sivak, B. L. Patritti, J. DiPisa, A. Caddle, K. Gilhooly, L. Govoni, S. Sivak, M. Lancia, R. Drillio and P. Bonato, "Patient specific ankle-foot orthoses using rapid prototyping," Journal of Neuroengineering and Rehabilitation, vol. 8, p. 1, January 2011.
19. S. Akinola and A. Telukdarie, "Sustainable digital transformation in healthcare: Advancing a digital vascular health innovation solution," Sustainability (Switzerland), vol. 15, 2023.
20. C. Alvarez-Romero, A. Martínez-García, M. Bernabeu-Wittel and C. L. Parra-Calderón, "Health data hubs: An analysis of existing data governance features for research," Health Research Policy and Systems, vol. 21, 2023.
21. H. A. Alzghaibi, "An examination of large-scale electronic health records implementation in primary healthcare centers in Saudi Arabia: A qualitative study," Frontiers in Public Health, vol. 11, 2023.
22. P. Apell and P. Hidefjäll, "Quality improvement: Understanding the adoption and diffusion of digital technologies related to surgical performance," International Journal of Quality and Reliability Management, vol. 39, pp. 1506–1529, 2022.
23. R. C. V. Loureiro, W. S. Harwin, K. Nagai and M. Johnson, "Advances in upper limb stroke rehabilitation: A technology push," Medical & Biological Engineering & Computing, vol. 49, pp. 1103–1118, 2011.
24. H. K. Channi, P. Shrivastava and C. L. Chowdhary, "Digital transformation in healthcare industry: A survey," Studies in Computational Intelligence, vol. 1039, pp. 279–293, 2022.
25. F. Dal Mas, M. Massaro, P. Rippa and G. Secundo, "The challenges of digital transformation in healthcare: An interdisciplinary literature review, framework, and future research agenda," Technovation, vol. 123, 2023.
26. D. Binci, G. Palozzi and F. Scafarto, "Toward digital transformation in healthcare: A framework for remote monitoring adoption," TQM Journal, vol. 34, pp. 1772–1799, 2022.
27. M. Breinbauer and M. Jansky, "Health apps in primary care: A waiting room survey in Rhineland–Palatinate [Gesundheits-apps in der hausärztlichen versorgung: Eine wartezimmerbefragung in rheinland-pfalz]," Pravention Und Gesundheitsforderung, 2023.
28. E. Cano-Marin, M. Mora-Cantallops and S. Sanchez-Alonso, "Prescriptive graph analytics on the digital transformation in healthcare through user-generated content," Annals of Operations Research: 1-25, 2023.
29. T. Engstrom, E. McCourt, M. Canning, K. Dekker, P. Voussoughi, O. Bennett, A. North, J. D. Pole, P. J. Donovan and C. Sullivan, "The impact of transition to a digital hospital on medication errors (TIME study)," NPJ Digital Medicine, vol. 6, 2023.
30. R. T. Tally Jr., "Snow crash," Dystopian States of America: Apocalyptic Visions and Warnings in Literature and Film, ABC-CLIO:2022, p. 281.
31. M. W. Bell, "Toward a definition of," Journal for Virtual Worlds Research, vol. 1, no. 1, 2008.

32. A. Bandyopadhyay, A. Sarkar, S. Swain, D. Banik, A. E. Hassanien, S. Mallik, A. Li and H. Qin, "A game-theoretic approach for rendering immersive experiences in the metaverse", *Mathematics*, vol. 11, no. 6, p. 1286, 2023.
33. A. Sihna, H. Raj, R. Das, A. Bandyopadhyay, S. Swain and S. Chakrborty, "Medical education system based on metaverse platform: A game theoretic approach", In IEEE 4th International Conference on Intelligent Engineering and Management (ICIEM 2023), pp. 1–6, 2023.
34. P. Gupta, K. Bhadani, A. Bandyopadhyay, D. Banik and S. Swain, "Impact of metaverse in the near 'future'", in IEEE 4th International Conference on Intelligent Engineering and Management (ICIEM 2023), pp. 1–6, 2023.
35. E. Z. Barsom, M. Graafland and M. P. Schijven, "Systematic review on the effectiveness of augmented reality applications in medical training," Surgical Endoscopy, vol. 30, pp. 4174–4183, October 2016.
36. S. K. Nayak, "Digital transformation roadmap: The case of Nova SBE's executive education," 2017.
37. R. B. Myerson, Game Theory: Analysis of Conflict, Harvard University Press, 1991.
38. B. Milovic and M. Milovic, "Prediction and decision making in health care using data mining," Kuwait Chapter of Arabian Journal of Business and Management Review, vol. 1, pp. 1–11, 2012.
39. E. J. Topol, "High-performance medicine: The convergence of human and artificial intelligence," Nature Medicine, vol. 25, pp. 44–56, January 2019.
40. M. M. T. Chit, W. Srisiri, A. Siritantikorn, N. Kongruttanachok and W. Benjapolakul, "Application of RFID and IoT technology into specimen logistic system in the healthcare sector," in 2021 3rd International Conference on Advancements in Computing (ICAC), 2021.
41. F. Jiang, Y. Jiang, H. Zhi, Y. Dong, H. Li, S. Ma, Y. Wang, Q. Dong, H. Shen and Y. Wang, "Artificial intelligence in healthcare: Past, present and future," Stroke and Vascular Neurology, vol. 2, pp. 230–243, December 2017.
42. J. E. Hollander and B. G. Carr, "Virtually perfect? Telemedicine for COVID-19," The New England Journal of Medicine, vol. 382, pp. 1679–1681, 2020.
43. J. Vamathevan, D. Clark, P. Czodrowski, I. Dunham, E. Ferran, G. Lee, B. Li, A. Madabhushi, P. Shah, M. Spitzer and S. Zhao, "Applications of machine learning in drug discovery and development," Nature Reviews Drug Discovery, vol. 18, pp. 463–477, 2019.
44. A. Alaiad and L. Zhou, "The determinants of home healthcare robots adoption: An empirical investigation," International Journal of Medical Informatics, vol. 83, pp. 825–840, 2014.
45. Y. Lu, "Industry 4.0: A survey on technologies, applications and open research issues," Journal of Industrial Information Integration, vol. 6, pp. 1–10, 2017.
46. W. Botta, V. de Donato, A. Persico and Pescapé, "Integration of cloud computing and Internet of Things: A survey," Future Generation Computer Systems, vol. 56, pp. 684–700, 2016.
47. P. P. Ray, "Home health hub Internet of Things (H3 IoT): An architectural framework for monitoring health of elderly people," in 2014 International Conference on Science Engineering and Management Research (ICSEMR), 2014.
48. Y. Yin, Y. Zeng, X. Chen and Y. Fan, "The Internet of Things in healthcare: An overview," Journal of Industrial Information Integration, vol. 1, pp. 3–13, 2016.
49. S. M. Riazul Islam, D. Kwak, M. Humaun Kabir, M. Hossain and K.-S. Kwak, "The Internet of Things for health care: A comprehensive survey," IEEE Access, vol. 3, pp. 678–708, 2015.
50. E. Dantas and R. Nogaroli, "The rise of robotics and artificial intelligence in healthcare: New challenges for the doctrine of informed consent," Medicine and Law, vol. 40, pp. 15–62, 2021.

51. N. Ganesh, R. Shankar, K. Kalita, P. Jangir, D. Oliva and M. Pérez-Cisneros, "A novel decomposition-based multi-objective symbiotic organism search optimization algorithm," Mathematics, vol. 11, p. 1898, 2023.
52. L. Zhang, Y. Luo, F. Tao, B. H. Li, L. Ren, X. Zhang, H. Guo, Y. Cheng, A. Hu and Y. Liu, "Cloud manufacturing: A new manufacturing paradigm," Enterprise Information Systems, vol. 8, pp. 167–187, 2014.
53. A. Wilson, H. Saeed, C. Pringle, I. Eleftheriou, P. A. Bromiley and A. Brass, "Artificial intelligence projects in healthcare: 10 practical tips for success in a clinical environment," BMJ Health and Care Informatics, vol. 28, 2021.
54. P. Galetsi, K. Katsaliaki and S. Kumar, "Assessing technology innovation of Mobile health apps for medical care providers," IEEE Transactions on Engineering Management, vol. 70, pp. 2809–2826, 2023.
55. K. Ghosh, M. S. Dohan, H. Veldandi and M. Garfield, "Digital transformation in healthcare: Insights on value creation," Journal of Computer Information Systems, vol. 63, pp. 449–459, 2023.
56. F. Tao, J. Cheng, Q. Qi, M. Zhang, H. Zhang and F. Sui, "Digital twin-driven product design, manufacturing and service with big data," The International Journal of Advanced Manufacturing Technology, vol. 94, pp. 3563–3576, February 2018.
57. A. I. Stoumpos, F. Kitsios and M. A. Talias, "Digital transformation in healthcare: Technology acceptance and its applications," International Journal of Environmental Research and Public Health, vol. 20, 2023.
58. D. Wu, J. L. Thames, D. W. Rosen and D. Schaefer, "Towards a cloud-based design and manufacturing paradigm: Looking backward, looking forward," in Volume 2: 32nd Computers and Information in Engineering Conference, Parts A and B, Chicago, 2012.
59. F. Tao, Y. Cheng, L. Zhang and A. Y. C. Nee, "Advanced manufacturing systems: Socialization characteristics and trends," Journal of Intelligent Manufacturing, vol. 28, pp. 1079–1094, 2017.
60. X. V. Wang and L. Wang, "From cloud manufacturing to cloud remanufacturing: A cloud-based approach for WEEE recovery," Manufacturing Letters, vol. 2, pp. 91–95, 2014.
61. S. Agrawal, S. Pandya, P. Jangir, K. Kalita and S. Chakraborty, "A multi-objective thermal exchange optimization model for solving optimal power flow problems in hybrid power systems," Decision Analytics Journal, vol. 8, p. 100299, 2023.
62. G. Shanmugasundar, V. Fegade, M. Mahdal and K. Kalita, "Optimization of variable stiffness joint in robot manipulator using a novel NSWOA-MARCOS approach," Processes, vol. 10, p. 1074, 2022.
63. G. Shanmugasundar, G. Sapkota, R. Čep and K. Kalita, "Application of MEREC in multi-criteria selection of optimal spray-painting robot," Processes, vol. 10, p. 1172, 2022.
64. V. Kumar, K. Kalita, P. Chatterjee, E. K. Zavadskas and S. Chakraborty, "A SWARA-CoCoSo-based approach for spray painting robot selection," Informatica, vol. 33, pp. 35–54, 2022.
65. G. Shanmugasundar, K. Kalita, R. Čep and J. S. Chohan, "decision models for selection of industrial robots—A comprehensive comparison of multi-criteria decision making," Processes, vol. 11, p. 1681, 2023.
66. S.-J. Baek and S.-H. Kim, "Robotics in general surgery: An evidence-based review," Asian Journal of Endoscopic Surgery, vol. 7, pp. 117–123, 2014.
67. L. A. McGuinness and B. Prasad Rai, "Robotics in urology," The Annals of The Royal College of Surgeons of England, vol. 100, pp. 45–54, 2018.
68. G.-Z. Yang, J. Bellingham, P. E. Dupont, P. Fischer, L. Floridi, R. Full, N. Jacobstein, V. Kumar, M. McNutt, R. Merrifield, B. J. Nelson, B. Scassellati, M. Taddeo, R. Taylor, M. Veloso, Z. L. Wang and R. Wood, "The grand challenges of science robotics," Science Robotics, vol. 3, p. eaar7650, January 2018.

69. R. Gailey, K. Allen, J. Castles, J. Kucharik and M. Roeder, "Review of secondary physical conditions associated with lower-limb amputation and long-term prosthesis use," Journal of Rehabilitation Research and Development, vol. 45, pp. 15–29, 2008.
70. S. Kurtz, K. Ong, E. Lau, F. Mowat and M. Halpern, "Projections of primary and revision hip and knee arthroplasty in the United States from 2005 to 2030," Journal of Bone and Joint Surgery, vol. 89, pp. 780–785, 2007.
71. P. Maceira-Elvira, T. Popa, A.-C. Schmid and F. C. Hummel, "Wearable technology in stroke rehabilitation: Towards improved diagnosis and treatment of upper-limb motor impairment," Journal of Neuroengineering and Rehabilitation, vol. 16, pp. 1–18, 2019.
72. J. Kim, A. S. Campbell, B. E.-F. de Ávila and J. Wang, "Wearable biosensors for healthcare monitoring," Nature Biotechnology, vol. 37, pp. 389–406, 2019.
73. B. D. Ratner, A. S. Hoffman, F. J. Schoen and J. E. Lemons, "Biomaterials science: An introduction to materials in medicine," MRS Bulletin, vol. 31, p. 59, 2006.
74. M. Sarraf, E. Rezvani Ghomi, S. Alipour, S. Ramakrishna and N. Liana Sukiman, "A state-of-the-art review of the fabrication and characteristics of titanium and its alloys for biomedical applications," Bio-Design and Manufacturing, p. 1–25, 2021.
75. K. Saranya, M. Rameez and A. Subramania, "Developments in conducting polymer based counter electrodes for dye-sensitized solar cells—An overview," European Polymer Journal, vol. 66, pp. 207–227, 2015.
76. A. Chortos, J. Liu and Z. Bao, "Pursuing prosthetic electronic skin," Nature Materials, vol. 15, pp. 937–950, September 2016.
77. R. M. Satava, "Historical review of surgical simulation–a personal perspective," World Journal of Surgery, vol. 32, pp. 141–148, February 2008.
78. B. Garrett, T. Taverner and P. McDade, "Virtual reality as an adjunct home therapy in chronic pain management: An exploratory study," JMIR Medical Informatics, vol. 5, p. e11, 2017.
79. S. Rajendran, N. Ganesh, R. Čep, R. C. Narayanan, S. Pal and K. Kalita, "A conceptual comparison of six nature-inspired metaheuristic algorithms in process optimization," Processes, vol. 10, p. 197, 2022.
80. W. Saad, Z. Han, H. Poor and T. Basar, "Game-theoretic methods for the smart grid: An overview of microgrid systems, demand-side management, and smart grid communications," IEEE Signal Processing Magazine, vol. 29, pp. 86–105, September 2012.
81. A. Mirhoseini, A. Goldie, H. Pham, B. Steiner, Q. V. Le and J. Dean, "A hierarchical model for device placement," in International Conference on Learning Representations, 2018.
82. A. Frisinger and P. Papachristou, "The voice of healthcare: Introducing digital decision support systems into clinical practice—A qualitative study," BMC Primary Care, vol. 24, 2023.
83. M. Joshi, K. Kalita, P. Jangir, I. Ahmadianfar and S. Chakraborty, "A conceptual comparison of dragonfly algorithm variants for CEC-2021 global optimization problems," Arabian Journal for Science and Engineering, vol. 48, pp. 1563–1593, 2023.
84. K. Shaik, J. V. N. Ramesh, M. Mahdal, M. Z. U. Rahman, S. Khasim and K. Kalita, "Big data analytics framework using squirrel search optimized gradient boosted decision tree for heart disease diagnosis," Applied Sciences, vol. 13, p. 5236, 2023.
85. M. Sulaiman, A. Håkansson and R. Karlsen, "AI-enabled proactive mHealth: A review," Communications in Computer and Information Science, vol. 1538, pp. 94–108, 2021.
86. H. M. Krumholz, "The time for digital health is almost here," Yonsei Medical Journal, vol. 63, pp. 493–498, 2022.
87. S. K. Mun, K. H. Wong, S.-C. B. Lo, Y. Li and S. Bayarsaikhan, "Artificial intelligence for the future radiology diagnostic service," Frontiers in Molecular Biosciences, vol. 7, 2021.

88. F. Kitsios and N. Kapetaneas, "Digital transformation in healthcare 4.0: Critical factors for business intelligence systems," Information (Switzerland), vol. 13, 2022.
89. S. Kraus, F. Schiavone, A. Pluzhnikova and A. C. Invernizzi, "Digital transformation in healthcare: Analyzing the current state-of-research," Journal of Business Research, vol. 123, pp. 557–567, 2021.
90. A. Hasselgren, K. Kralevska, D. Gligoroski, S. A. Pedersen and A. Faxvaag, "Blockchain in healthcare and health sciences—A scoping review," International Journal of Medical Informatics, vol. 134, p. 104040, 2020.
91. A. Dubovitskaya, Z. Xu, S. Ryu, M. Schumacher and F. Wang, "Secure and trustable electronic medical records sharing using blockchain," in AMIA Annual Symposium Proceedings, 2017.
92. M. V. Perez, K. W. Mahaffey, H. Hedlin, J. S. Rumsfeld, A. Garcia, T. Ferris, V. Balasubramanian, A. M. Russo, A. Rajmane, L. Cheung, G. Hung, J. Lee, P. Kowey, N. Talati, D. Nag, S. E. Gummidipundi, A. Beatty, M. T. Hills, S. Desai, C. B. Granger, M. Desai and M. P. Turakhia, A.H.S. Investigators, "Large-scale assessment of a smartwatch to identify atrial fibrillation," The New England Journal of Medicine, vol. 381, pp. 1909–1917, November 2019.
93. A. Azaria, A. Ekblaw, T. Vieira and A. Lippman, "MedRec: Using blockchain for medical data access and permission management," in 2016 2nd International Conference on Open and Big Data (OBD), Vienna, 2016.
94. K. Raghunathan, L. McKenna and M. Peddle, "Informatics competency measurement instruments for nursing students: A rapid review," Computers Informatics Nursing, vol. 40, pp. 466–477, 2022.
95. J. Priyadarshini, M. Premalatha, R. Čep, M. Jayasudha and K. Kalita, "Analyzing physics-inspired metaheuristic algorithms in feature selection with k-nearest-neighbor," Applied Sciences, vol. 13, p. 906, 2023.
96. X. Su and S. Zenios, "Patient choice in kidney allocation: The role of the queueing discipline," Manufacturing & Service Operations Management, vol. 6, pp. 280–301, October 2004.
97. A. L. Trejos, R. V. Patel and M. D. Naish, "Force sensing and its application in minimally invasive surgery and therapy: A survey," Proceedings of the Institution of Engineers, Part C: Journal of Mechanical Engineering Science, vol. 224, pp. 1435–1454, 2010.
98. R. H. Weber, "Internet of Things—New security and privacy challenges," Computer Law & Security Review, vol. 26, pp. 23–30, 2010.
99. C.-W. Tsai, C.-F. Lai and A. V. Vasilakos, "Future Internet of Things: Open issues and challenges," Wireless Networks, vol. 20, pp. 2201–2217, 2014.
100. R. C. Narayanan, N. Ganesh, R. Čep, P. Jangir, J. S. Chohan and K. Kalita, "A novel many-objective Sine–Cosine algorithm (MaOSCA) for engineering applications," Mathematics, vol. 11, p. 2301, 2023.
101. N. Raimo, I. De Turi, F. Albergo and F. Vitolla, "The drivers of the digital transformation in the healthcare industry: An empirical analysis in Italian hospitals," Technovation, vol. 121, 2023.
102. E. Tjoa and C. Guan, "A survey on explainable artificial intelligence (Xai): Toward medical Xai," IEEE Transactions on Neural Networks and Learning Systems, vol. 32, pp. 4793–4813, 2020.
103. A. Ekblaw, A. Azaria, J. D. Halamka and A. Lippman, "A Case Study for Blockchain in Healthcare: "MedRec" prototype for electronic health records and medical research data," in Proceedings of IEEE Open & Big Data Conference, 2016.
104. P. Lampreave, G. Jimenez-Perez, I. Sanz, A. Gomez and O. Camara, "Towards assisted electrocardiogram interpretation using an AI-enabled augmented reality headset," Computer Methods in Biomechanics and Biomedical Engineering: Imaging and Visualization, vol. 9, pp. 349–356, 2021.

105. F. Schiavone, G. Rivieccio, D. Leone, A. Caporuscio, M. Pietronudo, A. Bastone, M. Grimaldi, E. Celentano, A. Crispo, D. D'Errico, E. Coppola, F. Nocerino, S. Pignata and A. Bianchi, "Structuring the basis for performance measurement in a cancer network. An explorative analysis," Journal of General Management, vol. 48, pp. 386–393, 2023.
106. J. Padikkapparambil, C. Ncube, K. K. Singh and A. Singh, "Internet of Things Technologies for Elderly Health-Care Applications," in Emergence of Pharmaceutical Industry Growth With Industrial IoT Approach, Elsevier, 2020, pp. 217–243.
107. N. Ganesh, R. Shankar, R. Čep, S. Chakraborty and K. Kalita, "Efficient feature selection using weighted superposition attraction optimization algorithm," Applied Sciences, vol. 13, p. 3223, 2023.
108. K. Kalita and S. Chakraborty, "An efficient approach for metaheuristic-based optimization of composite laminates using genetic programming," International Journal on Interactive Design and Manufacturing (IJIDeM), vol. 17, pp. 899–916, 2023.
109. Z. Merchant, E. T. Goetz, L. Cifuentes, W. Keeney-Kennicutt and T. J. Davis, "Effectiveness of virtual reality-based instruction on students' learning outcomes in K-12 and higher education: A meta-analysis," Computers & Education, vol. 70, pp. 29–40, January 2014.

3 Metaverse

The Next Version of the Internet

Parijat Chatterjee
School of Computer Science and Engineering, Kalinga Institute of Industrial Technology, Bhubaneshwar, India

Devansh Adwani
School of Computer Science and Engineering, Kalinga Institute of Industrial Technology, Bhubaneshwar, India

Anjan Bandyopadhyay
School of Computer Science and Engineering, Kalinga Institute of Industrial Technology, Bhubaneshwar, India

Bhaswati Sahoo
School of Computer Science and Engineering, Kalinga Institute of Industrial Technology, Bhubaneshwar, India

3.1 INTRODUCTION

The Metaverse is a digital environment that unifies physical, augmented, and virtual reality (VR) within a communal online space. The alignment of these distinct realities results in a cohesive environment that can be perceived by all users from their unique vantage points. The technologies of VR, augmented reality (AR), and mixed reality (MR) are integral constituents of the metaverse, and their technologies exhibit significant overlap.

The present scholarly work examines the Metaverse and its capacity to influence various industries, including healthcare, manufacturing, and digital tools. The text delves into various aspects of the Metaverse, including its technological trajectory, identity modeling, decentralized technology, and social computing, shown in Figure 3.1.

3.1.1 The Technological Roadmap of the Metaverse

The manufacturing industry has undergone significant transformation through the utilization of Internet of Things (IoT) technology and the Metaverse core stream. According to a recent survey, a significant proportion (43%) of manufacturing firms

DOI: 10.1201/9781003449256-3

FIGURE 3.1 Visualization in the Metaverse.

anticipate the integration of VR technology into their operations to become commonplace within the next two to three years. The Metaverse is a forthcoming iteration of the internet, and numerous corporations are presently utilizing 5G and 6G technologies to construct their framework. The platform provides a simulated environment where individuals can engage with three-dimensional digital entities and virtual representations of themselves in a sophisticated manner that emulates real-life interactions. Metaverse services can facilitate access to services that are hosted in decentralized edge computing deployments, as well as access to services in the pre-existing cloud environment where all services are provided by centralized servers via the internet.

3.1.2 The Utilization of Digital Tools in the Metaverse

Digital humans are three-dimensional representations of chatbots that are present within the Metaverse and play a crucial role in shaping the overall terrain of the Metaverse. Artificial intelligence (AI) can analyze and convert natural languages into machine-readable formats, process the data, generate a response, and then convert the results back into English for communication with the user. Spatiotemporal AI refers to a set of computational resources, theoretical frameworks, and analytical techniques that can be utilized to enhance comprehension of the spatial and temporal dynamics of human behavior in virtual urban environments within the Metaverse. Utilizing AI techniques, the Metaverse manages distinct personal data-processing procedures in various countries.

The technologies of the metaverse can be classified into two primary domains: the technology that facilitates the metaverse and its associated ecosystem, and the ecosystem per se. The metaverse is reliant on key technological components such as edge computing and cloud computing, which are integral to its operation as a vast application. The utilization of edge computing has been shown to enhance the efficiency of applications that require low latency and high bandwidth, whereas cloud computing offers exceptional scalability in terms of computational resources and

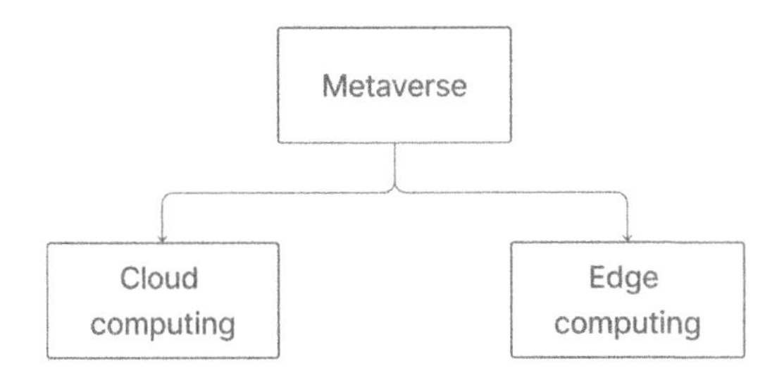

FIGURE 3.2 Technologies used in the Metaverse.

storage capabilities. Furthermore, the integration of edge devices and cloud services with sophisticated mobile networks has the potential to facilitate a diverse range of Metaverse technologies, including computer vision, AI, robots, and the IoT, contingent upon the availability of suitable hardware infrastructure, shown in Figure 3.2.

The Metaverse's ecosystem comprises various components, such as avatars that symbolize human users, content generation, the virtual economy, social conventions and regulations, and the significant challenges associated with constructing the Metaverse. The concept of the Metaverse involves the exchange of information and manipulation of objects, and multiple users within a virtual environment must receive uniform information and engage with one another coherently and instantaneously. The alignment of social norms and regulations with crucial factors such as social acceptability, security, privacy, trust, and accountability is imperative.

The academic and industry sectors place significant emphasis on four main categories of realities, namely, VR, AR, MR, and holographic technologies.

3.2 VIRTUAL REALITY

VR represents the extreme end of the reality–virtuality continuum, requiring users equipped with VR headsets to fully immerse themselves in virtual environments and disengage from physical reality, shown in Figure 3.3.

The technology of immersive VR is considered to be a significant emerging field in the 21st century and has garnered significant interest from consumers, practitioners, and scholars alike. The field of information systems (IS) has historically overlooked immersive VR technology. However, recent advancements in this technology have opened up new prospects for both organizations and IS scholars. Our study conducted a comprehensive analysis of the existing literature on immersive VR, intending to present a comprehensive perspective on the potential benefits and obstacles that organizations may encounter in this domain. Additionally, we identified potential avenues for future research in the field of IS. The study revealed the presence of five distinct affordances, namely, embodiment, interactivity, navigability, sense-ability, and create-ability. This chapter conducted a comprehensive review of 151 studies from the IS and related fields. The aim of this review was to synthesize the ways in which these affordances were utilized across various research domains.

The strategic opportunities and challenges associated with the implementation of VR were also identified. The implementation of immersive VR capabilities within

FIGURE 3.3 Interacting with the Metaverse.

organizations presents a promising field of study for scholars in the IS discipline. This area offers numerous opportunities to advance both the IS field and research on VR in organizational contexts.

The scope of VR research in conventional IS journals has been restricted, with a predominant emphasis on non-immersive VR, specifically desktop-based VR. This chapter provides a synthesis of 151 studies pertaining to immersive VR across 5 primary research domains, namely, training and education, healthcare, service and retail industries, management and organizing, and human–computer interaction. This review establishes a basis and outlines potential avenues for research on diverse strategic implementations of this nascent technology within organizational contexts. The perspective of technology affordance is regarded as a proficient functional framework for amalgamating the research works in our comprehensive review of the literature.

3.3 AUGMENTED REALITY

The concept of AR holds significant importance in the realm of digital interaction, and it encompasses a diverse range of forms and applications. As a result, it presents a plethora of intriguing research opportunities. The Metaverse is a vast 3D networked virtual environment that has the capacity to accommodate a significant number of individuals concurrently, facilitating social interaction. The proposition posits the emergence of a novel category of enhanced social engagement, denoted as "augmented duality." The concept of Metaverses offers a broader virtual domain for integration with the physical world. A novel research endeavor, conducted jointly by the Georgia Institute of Technology and Ludwig-Maximilians Universität Munich, has explored the incorporation of an AR element into Second Life [14–18]. The utilization of creativity facilitates the generation of novel varieties of interactive digital media, thereby bringing about a transformation in the manner in which individuals interact with the Metaverse. This, in turn, leads to the emergence of three augmentations pertaining to the Metaverse and the real world.

The paramount notion is that emphasizing positivity over negativity and directing attention toward constructive aspects rather than negative ones is crucial. This chapter explores the potential for interaction between the physical world and the Metaverse through the application of image-processing techniques such as background subtraction and overlays, and augmented duality is a novel form of AR interaction that facilitates social engagement through the presence of multiple contacts and parallel existences. Spellbinder is a software application that operates on a server and is designed for use with mobile devices equipped with cameras. Its primary function is to integrate and disseminate digital media, including text, audio, visual images, and video, from real-world objects. The phenomenon suggests that tangible reality and hitherto impartial entities are imbued with a sense of responsiveness as they assume the role of interactive digital media.

The spellbinder technology employs visual representations to generate gateways that connect the physical realm to the virtual world, enabling individuals to establish hyperlinks between perspectives of sites within Second Life and tangible entities or locations in reality. An MR boundary, which was symmetric and large in size, was established between two rooms that overlapped in both the physical world and Second Life. This boundary facilitated interaction between avatars and individuals in the physical world, as they were able to engage with one another by peering through the expansive screen located within their respective spaces. In Second Life, both avatars and human users have the ability to experience AR and augmented virtuality. This is achieved through the utilization of a large screen MR boundary within a physical space, which is then combined with either a virtual or real background within the virtual world. This chapter revealed that the system was perceived as both engaging and entertaining, albeit with certain constraints, such as the need to maintain visual focus on the screen and the potential for onlookers to observe the avatars. The design team made a decision to enhance the representation of human presence in Second Life by utilizing 3D video sprites. These sprites manifest as a video sprite within the space, as opposed to an overlay on a screen.

The design team executed three design interventions aimed at superimposing a Metaverse onto the physical world through the utilization of camera phones, image processing, and AR techniques. The outcome of these interventions was a proof of concept that featured a solitary user and avatar. The advent of Metaverse has led to the emergence of a novel form of augmented interaction, akin to the evolution of social networking from traditional web applications. Metaverse augmentations refer to a novel category of augmented interactions that facilitate intricate networks of concurrent social presence in both the physical world and the Metaverse. AR provides users with various sensory inputs, including audio, visual, olfactory, and tactile cues, to enhance their experiences within their immediate physical environment [19]. AR facilitates effortless and unobtrusive user engagement with digital entities in AR, thereby connecting human users in the physical world with the Metaverse. This process necessitates considerable endeavors in detection and tracking to accurately align virtual contents with their corresponding positions in the real environment. AR headsets offer distinct benefits compared to alternative methods with regard to facilitating user attention shifts and freeing up hands, thereby enabling users to effortlessly engage with the metaverse via an AR interface.

3.4 MIXED REALITY

MR lies between AR and VR, blending real and digital worlds for enhanced interaction. Existing as one of six definitions, MR objects demonstrate compatibility between digital and physical entities. Often considered an advanced AR, MR contributes to the Metaverse's creation through digital twins and user-generated content.

MR technology requires specialized hardware, like smart glasses, for realistic augmentation of reality. Microsoft's Hololens 2 exemplifies advancements in this field. AR, VR, and MR platforms digitize space for human experiences with distinct features, but a shared goal. Generation Z consumers may avoid over-reliance on AR, preferring personal perception.

Milgram and Kishino [20] introduced Metaverses as mixed-reality environments with continuous coexistence of real and virtual attributes. Metaverses lie within MR's spectrum, offering new opportunities for virtual environment design and usability. The Metaverse roadmap outlines strategic steps for its development, with a framework based on virtual simulation versus AR, and individual identity to collective world-building experiences.

Metaverse [21–23] technologies include AR, lifelogging, mirror worlds, and virtual worlds. AR is popular on social media platforms like Instagram and TikTok, but user-generated content creation remains limited. Lifelogging gains traction through wearables, while mirror worlds provide virtualized models of the physical world. Virtual worlds, like Fortnite or Roblox, let users build and engage in 3D settings.

3.5 IDENTITY MODELING, DECENTRALIZED TECHNOLOGY, AND SOCIAL COMPUTING

The Metaverse is a platform that integrates identity modeling, decentralized technology, and social computing to create a VR experience. The process of identity modeling pertains to the development of a distinct three-dimensional representation, commonly referred to as an avatar, within virtual environments. Decentralized technology encompasses a wide range of areas, such as land, space, energy, power, buildings, medical healthcare materials, and others. The storage of data is completely decentralized, with ownership and management distributed among all parties involved, thereby ensuring its persistence. The blockchain serves as the fundamental basis for the infrastructure, guaranteeing the trustworthiness of decentralized data, decentralized databases, and decentralized computation. This ensures exclusive ownership of all virtual assets within the Metaverse by its citizens. The field of social computing encompasses various aspects, such as the depiction of avatars, the identification of avatars, the interaction between avatars, and the utilization of avatars for organizational purposes.

The topic of interest pertains to the domains of extended reality (XR) and feature extraction.

The management of Metaverse XR involves the utilization of AI methodologies. VR, AR, and MR are three distinct technological advancements that integrate the physical and virtual realms. The process of feature extraction entails identifying and isolating distinct attributes and patterns that serve as indicators for the existence or non-existence of a specific intention. The process of feature interpretation entails scrutinizing the attributes acquired in the preceding stages and drawing a deduction regarding the possible intention of the user. There are multiple manifestations of output, such as visual displays, cursor movements on a screen, or adjustments to audio volume.

3.6 HEALTHCARE SECTOR

Metaverse technology offers potential advantages in healthcare, enabling remote communication and collaboration between practitioners and patients and enhancing medical instruction and learning [1]. This chapter examines its potential applications in healthcare, given in Table 3.1.

Metaverse technology can provide immersive experiences in healthcare education, fostering problem-solving skills and immediate feedback through AR. Virtual classrooms, study rooms, and meetings can be created for effective learning.

Surgical training: Metaverse technology can offer immersive training experiences for various surgical procedures. Digital surgery provides AI-powered surgical video management and digital solutions for hospitals [2]. VR simulators have been developed for dental anesthetic training and submandibular glandectomy with endoscopic assistance.

Clinical care: Metaverse technology can enhance clinical care experiences. Augmedics headset technology has been used for AR surgeries, AccuVein for vein

TABLE 3.1
Applications of Metaverse in Healthcare Sector

Application	Description
Surgical training	Facilitates immersive training in surgical procedures, powered by AI and VR simulators
Clinical care	Augmented reality headsets project internal anatomy during surgeries, aiding in intravenous injections
Pain management	VR therapy diverts patients' attention from pain, resulting in a 50% reduction compared to conventional methods
Chronic diseases	Immersive experiences educate patients on their conditions, aiding in symptom management
Telemedicine	Enhances remote healthcare access with immersive encounters, replicating in-person interactions
Simulation training	Creates life-like simulations for medical education, improving engagement and immersion
Medical research	Provides 3D models of organs and tissues for examination, facilitating new therapies and interventions

visualization, and virtual environments have been created for obstetrics to mitigate pain signals during pregnancy. Pregnant women in NHS hospitals use VR headsets to relax in simulated environments. Eye-Sync, a VR headgear by SyncThink, demonstrates metaverse application in neurology with infrared cameras observing ocular motion. The Cleveland Clinic utilizes metaverse technology to improve liver cancer treatment.

This chapter aims to examine case studies that demonstrate the potential applications of Metaverse technology within the healthcare industry.

1. VR for pain management: VR therapy helps alleviate pain, anxiety, and other medical conditions [3]. For burn injury patients, it reduces pain by 50% compared to conventional methods (University of Washington study).
2. Metaverse for chronic disease management: Metaverse technology facilitates patient education and symptom management. Onduo created a virtual diabetes clinic with AI and Metaverse technology for individualized coaching and assistance.
3. Metaverse in telemedicine: Metaverse technology enhances telemedicine with immersive experiences replicating in-person interactions. XRHealth created a telehealth platform using VR and AR for remote physical therapy, mental health counseling, and other services.
4. Simulation-based training: Metaverse technology generates lifelike simulations, improving engagement and immersion compared to conventional techniques. Osso VR created a VR training platform for simulating surgical procedures digitally.
5. Metaverse in medical research: 3D Systems developed a virtual anatomy platform for creating 3D organ and tissue models, facilitating novel therapies and interventions for various illnesses.

These case studies demonstrate Metaverse technology applications in healthcare, including pain management, chronic disease management, medical education, telemedicine, and research [4].

Metaverse technology holds promise for revolutionizing healthcare delivery and reception through immersive experiences and efficiency improvements.

3.7 EDUCATION SECTOR

The Metaverse is a virtual environment that offers users a fully immersive experience, enabling them to engage with each other and the surroundings in real-time. The potential applications of this technology in various fields of education, such as medical, nursing, healthcare, science, military, and manufacturing training, are noteworthy [5]. This chapter investigates the utilization of the Metaverse in the domain of education and proposes potential avenues for its integration in the field, given in Table 3.2.

Metaverse in education: The Metaverse enables knowledge and skill acquisition through interaction with tutors, peers, and students in an immersive living environment [6]. It offers numerous educational applications, facilitating collaboration, peer interaction, and work-related activities.

TABLE 3.2
Applications of Metaverse in the Education Sector

Applications	Description
Immersive learning	Facilitates knowledge and skill acquisition through interaction with tutors, peers, and students; simulates an immersive living environment
Reasons for adoption	Offers continuous, safe skill-practicing settings; motivates learners; provides alternative perspectives and roles for career or life-related tasks
Research topics	Formulation of Metaverse-based educational models/frameworks; establishment of institutions and practice spaces; comprehensive training programs and certification

Reasons for adoption: The Metaverse provides a continuous, safe skill-practicing setting, motivates learners, and allows exploration of potential through complex, diverse, and authentic tasks.

Research topics: Research areas include developing Metaverse-based educational models, establishing institutions and practice spaces, and providing comprehensive training programs and certification. Scholars investigate the impact of metaverse-based education on academic achievements and attitudes.

Concept mapping and inquiry-based learning: Metaverse concept mapping may differ from its physical application, affecting inquiry-based learning [7]. The system can assist users, allowing them to make genuine decisions in authentic scenarios, unlike traditional business management courses.

The Metaverse as a novel learning environment: An inquiry into the perceptions of learners with varying personal characteristics regarding the Metaverse as an emerging educational concept [8], and its potential benefits for their engagement in novel contexts, is warranted.

1. University of Cincinnati: UCSIM3, a Metaverse learning platform, focuses on healthcare and bioengineering, providing immersive experiences and enhancing educational efficacy.
2. SK Telecom: Jump VR explores Metaverse implementation in academia, targeting linguistic acquisition and cultural immersion to improve academic achievements and create novel learning opportunities.
3. Stanford University: Virtual human course investigates identity, empathy, and social interaction, promoting experiential learning, academic retention, and immersive educational opportunities.
4. Embry-Riddle Aeronautical University: Developing a virtual collision laboratory, aiming to provide immersive experiences in a secure setting and become a prominent Metaverse academic center.
5. Case Western Reserve University: Incorporates Metaverse technology using Microsoft's "Hololens" to enhance human anatomy and physiology understanding, aiming to improve academic achievements and address medical complexities.

To summarize, the metaverse holds considerable promise for incorporation within the realm of education, providing learners with immersive, authentic, and interactive learning environments. Prospective investigations in this domain ought to concentrate on the advancement of educational models and execution frameworks based on Metaverse technology, as well as the provision of comprehensive training programs and certificates to learners. Moreover, it is imperative to conduct further research on the impact of Metaverse-based educational settings on the academic achievements and perspectives of learners.

3.8 BANKING SECTOR

The Metaverse platform employs blockchain technology to revolutionize the banking sector through the provision of secure and uninterrupted transactions, enhanced customer communication, and support for carbon neutrality objectives [12]. The objective of this chapter is to investigate the potential ramifications of Metaverse on the banking sector, and its capacity to stimulate innovation and enhance efficacy in diverse domains, including customer communication, digital identity, mortgages, and NFTs, given in Table 3.3.

Customer communication: Metaverse emphasizes customer communication, offering a streamlined approach for managing account information. It integrates banks with other financial systems, reducing costs and improving efficiency. By harnessing AI and seamlessly integrating with VR platforms, Metaverse cultivates robust connections with clients and accommodates customers with limited physical access due to travel or busy schedules.

Digital identity: The Metaverse platform uses smart contracts to establish clear ownership and transaction history, providing a comprehensive digital identity verification system for clients. This system eliminates the need for manual KYC processes, saving time and costs while reducing each company's carbon footprint. It addresses the complexities of opening accounts or fulfilling KYC requirements in various jurisdictions.

TABLE 3.3
Applications of Metaverse in Healthcare Sector

Application	Description
Customer communication	Enhances communication through streamlined data management, VR integration, and AI-based messaging
Digital identity	Offers a digital identity verification system with smart contracts, simplifying KYC processes and reducing carbon footprint
Mortgages	Revolutionizes mortgage system using blockchain technology for transparency, equity, and minimized fraud and loan defaults
NFTs	Provides flexibility in transactions, supports credit card processing, mortgage lending, and showcases environmental commitment
Carbon net zero goals	Contributes to carbon neutrality through cost reduction in the energy sector and reliable accounting using blockchain technology

Mortgages: Metaverse employs blockchain technology to revolutionize the mortgage system, allowing lenders to quickly evaluate borrowers' creditworthiness based on factors like income history, job tenure, credit rating, and assets. Smart contracts on Metaverse ensure transparency and fairness in loan agreements, mitigating fraudulent activities and minimizing loan defaults.

Non-fungible tokens: Non-fungible tokens (NFTs) provide financial institutions with enhanced flexibility in conducting transactions, bank account management, credit card processing, and mortgage lending. NFTs can be applied to credit cards and property-based lending, showcasing institutions' commitment to environmental sustainability and fostering credibility among clients and stakeholders.

Carbon net zero goals: Metaverse implementation contributes to carbon net zero by reducing costs in the energy sector and providing a reliable accounting system. Its blockchain technology guarantees transparency and accountability, eliminating obstacles arising from diverse currencies and intricate banking systems in cross-national financial transactions.

Case studies: HSBC's employment of Metaverse technology in the banking industry is exemplified by the implementation of "HSBC Island" in Second Life, a virtual environment that enables HSBC clients to acquire knowledge about the bank's offerings and solutions [13]. Within this digital realm, clients possess the ability to engage with one another and financial institution agents in a synchronous manner. Swiss Bank Sygnum employs NFTs to facilitate the fractional ownership of tangible assets, such as real estate, artwork, and collectables, by issuing asset-backed NFTs. Additionally, the financial institution has initiated its operations.

3.9 BLOCKCHAIN TECHNOLOGY

The concept of Metaverse has been in existence for nearly three decades and has recently garnered attention owing to advancements in diverse technologies such as blockchain, IoT, VR/AR, AI, and cloud/edge computing, among others [9]. The emergence of these technological advancements has brought about the realization of the Metaverse, with prominent companies such as Facebook outlining crucial attributes including identity, social connections, immersive engagement, minimal obstacles, respectful conduct, economic viability, universal accessibility, and diverse offerings. The Metaverse encounters several obstacles stemming from its centralized economic structure and substantial transactional volumes. As a result, the implementation of blockchain and AI technologies has been investigated as prospective remedies to surmount these obstacles, given in Table 3.4.

Digital creation: The Metaverse is built upon digital creation, which necessitates a fundamental operational tool for seamless personalization. Decentraland offers a range of resources for the development of immersive applications and games with interactive features.

Digital creation: The Metaverse relies on digital creation for crafting personalized experiences, with platforms like decentraland offering resources to develop immersive applications, games, and interactive experiences that enrich user engagement.

Digital asset: Digital assets, such as weapon enhancements or "skins," hold inherent value that users can trade, exchange, or purchase. However, trading platforms

TABLE 3.4
Applications of Metaverse in Blockchain Technology

Application	Description
Machine learning	Algorithms that learn from experience and data, impacting metaverse efficiency and cognition
Proposed frameworks	Novel virtual environments for visual deep learning, AR devices, and federated learning
Data processing	Blockchain technology reconciles simulated/real-world differences and eases Metaverse load
Transmission and verification	Supports diverse data transmission and validation in metaverse economic systems
Distributed storage	Ensures security of virtual assets and user identities through blockchain technology
Consensus mechanism	PoW and PoS mechanisms used for consensus, with PoS focused on equitable miner participation
Smart contract	Facilitates value exchange and transparent execution of system regulations
Cryptocurrency	Digital currency creation, recording, and trading essential for the Metaverse
Digital creation	Decentraland provides resources for immersive applications and games in the Metaverse
Digital asset	Tradeable enhancements with potential privacy concerns on public trading platforms
Digital market	Platform for commercial transactions generating revenue in the physical world

that publicly disclose user accounts may raise privacy concerns and potential misuse of information.

Digital market: The digital market serves as a key platform for commercial transactions that generate real-world revenue. For the Metaverse to mature, it is essential to evolve beyond the current digital market model and explore new paradigms.

Blockchain and AI: Blockchain and AI technologies offer solutions to Metaverse challenges, such as large dataset analysis, AI-driven content generation, and deployment of intelligence. They also play a crucial role in protecting digital assets, currencies, and markets within the Metaverse ecosystem.

Machine learning: Machine learning algorithms enable learning from experience and data, with techniques like collective-weight frameworks and convolutional kernels having the potential to greatly enhance the Metaverse's operational efficiency and cognitive capabilities.

Proposed frameworks: Researchers are proposing novel virtual environments for visual deep learning, AR devices, and federated learning. These frameworks facilitate efficient data-sharing and insights gathering for various applications, such as emergency response training.

Data processing: Blockchain technology can help reconcile differences between simulated and real-world environments, reduce computational load in the Metaverse, and support essential functions of digital currencies, fostering seamless integration and growth.

Transmission and verification mechanisms: Data transmission and verification mechanisms provide network support for diverse data transmission and validation of economic systems in the Metaverse. These mechanisms, coupled with smart contracts, ensure transparency and compliance with system regulations.

Distributed storage: In the Metaverse, distributed storage in blockchain technology ensures the security of virtual assets and user identities. However, complete blockchain nodes are required to locally store all historical interactions, which can burden node owners.

Consensus mechanism: Bitcoin and Ethereum use the proof-of-work (PoW) consensus mechanism, while the proof-of-stake (PoS) mechanism leverages coin holdings to achieve consensus on proposed transactions. Adaptation within the Metaverse is crucial for equitable participation among miners.

Smart contract: Smart contract technology facilitates the exchange of value and transparent execution of system regulations. Decentralized exchanges enable cryptocurrency exchanges through smart contracts or peer-to-peer networks for automatic transaction execution.

Cryptocurrency: Cryptocurrencies like Bitcoin play a significant role in the Metaverse, with blockchain technology enabling operations such as creation, recording, and trading. These functions are crucial for seamless integration and expansion within the Metaverse ecosystem.

Presented below are several case studies [10] pertaining to the application of blockchain technology in the Metaverse.

1. Decentraland is a VR platform on the Ethereum blockchain, allowing users to create, experience, and monetize assets and applications, with community control over land ownership and creative activities.
2. Sandbox is a decentralized Metaverse on the Ethereum blockchain, enabling users to build and profit from gaming experiences using the native ERC20 compliant SAND token and create 3D entities authenticated as NFTs.
3. Axie infinity is a Play-to-Earn Metaverse project featuring fantasy creatures called Axies, with a blockchain-based economy that rewards players for advancing in-game skills and participating in governance activities using Axie Infinity Shards tokens.
4. Illuvium is a decentralized role-playing game Metaverse where players interact with creatures called Illuvials, represented by ERC-1155 tokens, with blockchain technology ensuring immutability and exploring new in-game economic models.

Each of these projects has leveraged blockchain technology to create a unique Metaverse experience for users.

To sum up, the implementation of blockchain technology within the metaverse has the potential to establish a decentralized and highly secure ecosystem for its users [11]. The implementation of blockchain-based solutions in the domains of professional certification, e-learning, and virtual and AR has the potential to augment user experience and generate novel prospects for content creators, educators, and learners. The global e-learning industry is anticipated to reach a valuation of

$325 billion by the year 2025. The integration of blockchain technology has the potential to establish a novel e-learning market within the metaverse.

REFERENCES

1. R. Harrill, Virtual reality significantly reduces pain-related brain activity, June 21, 2004, UW News. https://www.washington.edu/news/2004/06/21/virtual-reality-significantly-reduces-pain-related-brain-activity/
2. XRHealth, XRHealth Opens First Virtual Reality Telehealth Clinic, February 2020. Retrieved February 25, 2023, from https://www.prnewswire.com/news-releases/xrhealth-opens-first-virtual-reality-telehealth-clinic-301003283.html
3. 3DSystems Wikipedia, 3D Systems—Wikipedia. 3DSystems Wikipedia, January 1, 1986, https://en.wikipedia.org/wiki/3D_Systems
4. G. Wang, A. Badal, X. Jia, J. S. Maltz, K. Mueller, K. J. Myers, C. Niu, M. Vannier, P. Yan, Z. Yu and R. Zeng, "Development of metaverse for intelligent healthcare—Nature machine intelligence," Nature, vol. 4, 2022, pp. 922–929. https://doi.org/10.1038/s42256-022-00549-6
5. A. Tlili, R. Huang, B. Shehata, D. Liu, J. Zhao, A. H. Saleh Metwally, H. Wang, M. Denden, A. Bozkurt, L. H. Lee, D. Beyoglu, F. Altinay, R. C. Sharma, Z. Altinay, Z. Li, J. Liu, F. Ahmad, Y. Hu, S. Salha and D. Burgos, "Is Metaverse in Education a Blessing or a Curse: A Combined Content and Bibliometric Analysis," in Smart Learning Environments. SpringerOpen, July 6, 2022. https://doi.org/10.1186/s40561-022-00205-x
6. H. Lin, S. Wan, W. Gan, J. Chen and H. C. Chao, *Metaverse in Education: Vision, Opportunities, and Challenges*. arXiv.org, November 27, 2022. https://doi.org/10.48550/arXiv.2211.14951
7. G.J. Hwang, and C. Shu-Yun, "Definition, roles, and potential research issues of the metaverse in education: An artificial intelligence perspective," Computers and Education: Artificial Intelligence 3 (2022): 100082
8. J. Singh, M. Malhotra and N. Sharma, "Metaverse in education: An overview", in Metaverse in Education: An Overview: Business & Management, IGI Global. https://doi.org/10.4018/978-1-6684-6133-4.ch012
9. Q. Yang, Y. Zhao, H. Huang, Z. Xiong, J. Kang and Z. Zheng, "Fusing blockchain and AI with metaverse: A survey," IEEE Open Journal of the Computer Society, vol. 3, 122–136. https://doi.org/10.1109/ojcs.2022.3188249
10. D. Wei, "Gemiverse: The blockchain-based professional certification and tourism platform with its own ecosystem in the metaverse," International Journal of Geoheritage and Parks 10.2 (2022): 322–336.
11. B. Ryskeldiev, Y. Ochiai, M. Cohen and J. Herder, "Distributed Metaverse: creating decentralized blockchain-based model for peer-to-peer sharing of virtual spaces for mixed reality applications," in Proceedings of the 9th Augmented Human International Conference (AH '18), Association for Computing Machinery, New York, NY, USA, Article 39, 2018, pp. 1–3. https://doi.org/10.1145/3174910.3174952
12. V. Dubey, A. Mokashi, R. Pradhan, P. Gupta, and R. Walimbe, "Metaverse and banking industry—2023 the year of Metaverse adoption", Technium: Romanian Journal of Applied Sciences and Technology, 4(10), 62–73.
13. M. R. Hasta Anggara, M. R. Davie, M. Margani, D. M. A. Aristyana and M. Aulia. "The presence of commercial banks in Metaverse's financial ecosystem: Opportunities and risks." Journal of Central Banking Law and Institutions, 1(3), 405–430.
14. M. A. I. Mozumder, M. M. Sheeraz, A. Athar, S. Aich and H. C. Kim, "Overview: Technology roadmap of the future trend of Metaverse based on IoT, blockchain, AI technique, and medical domain Metaverse activity," in 2022 24th International

Conference on Advanced Communication Technology (ICACT), February 13, 2022. https://doi.org/10.23919/icact53585.2022.9728808

15. E. Dincelli, and Y. Alpe, "Immersive virtual reality in the age of the Metaverse: A hybrid-narrative review based on the technology affordance perspective," The Journal of Strategic Information Systems 31.2 (2022): 101717. https://doi.org/10.1016/j.jsis.2022.101717
16. M. Wright, H. Ekeus, R. Coyne, J. Stewart, P. Travlou and R. Williams, "Augmented duality," in Proceedings of the 2008 International Conference on Advances in Computer Entertainment Technology, December 3, 2008. https://doi.org/10.1145/1501750.1501812
17. D. Buhalis and N. Karatay, "Mixed reality (MR) for Generation Z in cultural heritage tourism towards Metaverse," in Information and Communication Technologies in Tourism 2022, 2022, pp. 16–27. https://doi.org/10.1007/978-3-030-94751-4_2
18. J. de la Fuente Prieto, L. Pilar, and R. Martínez-Borda, "Approaching Metaverses: Mixed reality interfaces in youth media platforms," New Techno Humanities 2.2 (2022): 136–145.. https://doi.org/10.1016/j.techum.2022.04.004
19. L. H. Lee, T. Braud, P. Zhou, L. Wang, D. Xu, Z. Lin, A. Kumar, C. Bermejo and P. Hui, "All one needs to know about Metaverse: A complete survey on technological singularity, virtual ecosystem, and research agenda," arXiv.org, 2021, October 6. https://doi.org/10.48550/arXiv.2110.05352
20. R. Skarbez, M. Smith and M. C. Whitton, "Revisiting Milgram and Kishino's reality-virtuality continuum. Frontiers, Frontiers in Virtual Reality 2 (2021): 647997. https://doi.org/10.3389/frvir.2021.647997
21. A. Bandyopadhyay, A. Sarkar, S. Swain, D. Banik, A. E. Hassanien, S. Mallik, A. Li and H. Qin, "A game-theoretic approach for rendering immersive experiences in the Metaverse", Mathematics, vol. 11, no. 6, 2022, p. 1286.
22. A. Sihna, H. Raj, R. Das, A. Bandyopadhyay, S. Swain and S. Chakrborty, "Medical education system based on Metaverse platform: A game theoretic approach", in IEEE 4th International Conference on Intelligent Engineering and Management (ICIEM 2023), pp. 1–6.
23. P. Gupta, K. Bhadani, A. Bandyopadhyay, D. Banik and S. Swain, "Impact of Metaverse in the near 'future'", in IEEE 4th International Conference on Intelligent Engineering and Management (ICIEM 2023), pp. 1–6.

4 Digital Twins in the Metaverse

Aryan Kaushal
School of Computer Science and Engineering, Kalinga Institute of Industrial Technology, Bhubaneshwar, India

Suchetan Mukherjee
School of Computer Science and Engineering, Kalinga Institute of Industrial Technology, Bhubaneshwar, India

Debarghya Roy
School of Computer Science and Engineering, Kalinga Institute of Industrial Technology, Bhubaneshwar, India

Anjan Bandyopadhyay
Department of Computer Science and Engineering, Kalinga Institute of Industrial Technology, Bhubaneshwar, India

Shivansh Mishra
School of Computer Science and Engineering, Kalinga Institute of Industrial Technology, Bhubaneshwar, India

Sujata Swain
School of Computer Science and Engineering, Kalinga Institute of Industrial Technology, Bhubaneshwar, India

4.1 INTRODUCTION

"The Digital Twin is one of the top concepts in product realization," proclaims Dr. Michael Grieves, the mind behind the very idea of Digital Twin technology [1]. Such an assertion underscores the monumental importance and transformative potential of Digital Twins (DTs), especially as we journey deeper into the vast expanses of the Metaverse. This ever-evolving digital cosmos, which spans areas as diverse as entertainment, healthcare, urban planning, and education, positions DTs at its vanguard, signaling a new age where physical and virtual realities converge.

At the outset, it is pertinent to delineate the very concept of a "Digital Twin." As a starting point, a DT is considered as a contextualized software model of a real-world object [2]. It, in its most fundamental sense, is a dynamic, virtual replica of a physical object, system, or process. It offers a real-time, holistic view by integrating

DOI: 10.1201/9781003449256-4

data from myriad sources and sensors, enabling stakeholders to test, visualize, and predict potential outcomes without physically interacting with the entity in question. This interplay of physical and digital realities, driven by the twin, paves the way for enhanced and innovative applications in the Metaverse – from simulating entire cities for disaster management to creating immersive educational experiences. The DT has found applications in the wide realm of the Internet of Things (IoT), providing a promising solution for expressing complex environments and sensing/actuation capabilities [3, 4]. For instance, the DT is utilized for the optimization of massive sensor networks and in the modeling of smart cities [5].

But, like every revolutionary technological leap, the use of DTs in the Metaverse is not without its set of challenges. On the one hand, they hold the promise of efficiency, safety, and adaptability, allowing organizations to pilot ideas, troubleshoot issues, and optimize solutions in a virtual arena before their real-world implementation. On the other hand, the complexities related to data privacy, the accuracy of replication, and the ethical considerations of creating and manipulating digital doppelgängers raise imperative questions.

The potential benefits of DTs are manifold. For individuals, they can personalize experiences in the Metaverse, fostering a deeper connection and immersion with virtual environments. For organizations, the ability to simulate, predict, and strategize in a risk-free digital space could mean significant cost savings, agility, and competitive advantage. On a broader societal level, DTs could revolutionize urban planning, healthcare delivery, and environmental conservation, to name just a few areas of impact.

However, the risks too are significant. The vast amount of data needed to create accurate DTs poses severe privacy concerns. Furthermore, the infrastructure required to gather, process, and represent the data may or may not exist [6]. There is also the threat of malicious actors manipulating Digital Twin data for nefarious purposes. Moreover, as the boundaries blur between what's real and what's virtual, psychological, ethical, and philosophical concerns might emerge about our relationship with our digital selves and the nature of reality itself in the Metaverse.

In this chapter, we embark on an exploratory journey into the realm of DTs and their symbiotic relationship with the Metaverse, unraveling the nuances of their applications, the challenges they pose, and the profound implications they hold for the future. As we navigate this intricate tapestry, the broader objective is to engender a nuanced understanding and foster a discourse on how we can harness this technology ethically and effectively.

4.2 CONCEPTS AND GENERAL CHALLENGES OF PRELIMINARIES

4.2.1 Digital Twin

DT is a virtual representation of a procedure, good, or service that was suggested by NASA in 1970 [6]. In the physical world, DTs give process prediction and risk prevention based on the input data [7]. The essential principle of deploying a digital duplicate to evaluate a physical object, however, can be seen considerably earlier. It is realistic to say that NASA was the first agency to deploy DT technology during its space exploration expedition [8].

The concept of DT remains elusive, with its definition being subjected to varied interpretations due to the surge in recent literature. Upon reviewing this expanding body of work, certain foundational technologies consistently emerge as pivotal in driving DT implementations. Among these, machine learning (ML) [9], transfer learning [10], and distributed computing paradigms – encompassing cloud, fog, and edge computing [11] – stand out. Additionally, the Industrial Internet of Things [12], cyber-physical systems [13], and virtual/augmented reality (VR/AR) [14] further underscore the technological backbone of DT.

4.2.1.1 Concept and Design of Digital Twins

DTs are virtual replicas of physical devices that data scientists and IT pros can use to perform simulations before actual devices are built and deployed. They are also altering the optimization of technologies like IoT, AI, and analytics. DTs are employed in many sectors of the real world, including industry, healthcare, and urban planning. DTs, for instance, can be used in manufacturing to track, analyze, and optimize production processes in real-time, lowering downtime and raising product quality. To simulate the effects of various treatment options and to forecast patient outcomes, DTs of patients can be generated in the healthcare industry.

The Digital Twin Model has three fundamental elements: First, it encompasses the tangible or envisioned physical entities, existent or destined to exist in the physical domain, referred to as the "Physical Twin." On the opposite side of the spectrum lies the virtual or digital equivalents that inhabit the virtual domain, properly called "the Digital Twin" [15].

This paradigm incorporates three key attributes (Figure 4.1): On one side, we keep the enduring presence of the physical space and tangible items that have persisted throughout history and continue to have a pivotal role in the real world. The requirement for actual, practical objects for real-world functionalities remains unshakable.

On the opposing side, we journey into the domain of virtual space – a digital simulation of the tangible products occupying the other side. Within this virtual domain, we encapsulate crucial information about these products.

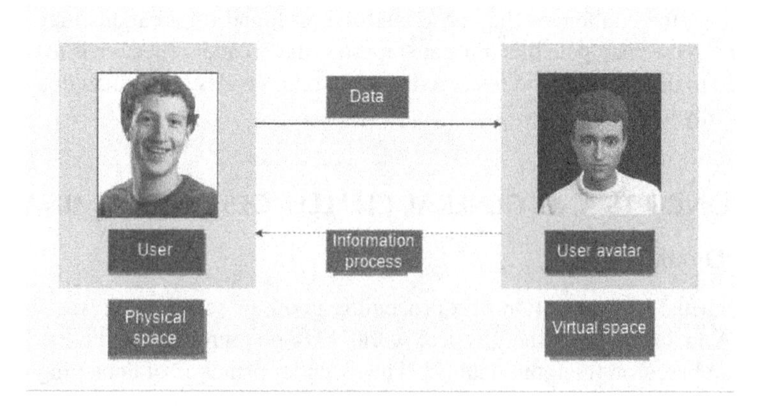

FIGURE 4.1 The contemporary model of the Digital Twin illustrates the alignment between the virtual and physical realms.

The third crucial component resides in developing a seamless connection between the actual and virtual spaces. This connection serves as the channel through which data are carried from the physical space into the virtual realm to develop and enrich our virtual products. Moreover, it promotes the usage of knowledge obtained from the virtual domain within the physical space. This connectivity between the two domains is sometimes referred to as the "Digital Thread."

4.2.1.2 General Challenges

The synergy between DT technology, artificial intelligence (AI), and IoT technology has become evident, bringing with it shared challenges. The initial stride toward resolution lies in pinpointing these challenges [16].

1. *IT infrastructure:* The burgeoning adoption of DT in the realms of Digital Analytics and the IoT highlights pressing IT infrastructure concerns. Key among these is the need for high-performance infrastructure, imperative for advanced algorithmic operations, especially with AI's rapid progression. Despite innovative solutions like "GPU as a service," cost and security concerns in cloud integrations remain prominent. Furthermore, the swift evolution of IoT poses integration dilemmas, especially with legacy systems. Without a robust and interconnected IT framework, achieving the full potential of DT, especially when intertwined with analytics and IoT, becomes considerably challenging.
2. *Data concerns:* In the sphere of Digital Analytics and DT, data quality arises as a critical challenge. For analytics, it's crucial to curate and improve the data, ensuring AI algorithms receive only premium, cleansed inputs. Similarly, a DT's efficacy strongly relies on continuous, noise-free, and high-quality data streams. Any breaches in data quality, defined by irregularities or interruptions, endanger the functioning of a DT. Consequently, thorough planning and perceptive analysis of IoT signals are needed to ensure the ideal data collection is exploited for a DT's effective operation.
3. *Privacy and security:* Data analytics, IoT, and DT are inextricably related to concerns of privacy and security. The rise of AI in analytics needs stronger laws, emphasized by General Data Protection Regulation in Europe. By decentralizing model training, innovations like federated learning alleviate DT privacy problems. Meanwhile, IoT's expansion magnifies risks, evidenced from occurrences like the Mirai botnet assault [17]. Ensuring security becomes critical, especially as DTs mix analytics and IoT, processing sensitive system data.
4. *Trust:* Both technological and perception-based issues influence how people perceive AI, IoT, and DT. AI's revolutionary but fledgling nature often provokes fear. To develop trust, the focus should move toward AI's beneficial outcomes, along with strict security standards. For IoT and DT, ensuring security and transparent operations is critical. DTs, in particular, require core understanding and frequent validation to strengthen trust from both businesses and end-users.

4.2.2 Metaverse

Metaverse is a hypothetical artificial world related to the physical world, defined by the prefix "meta" (implying transcending) and the word "universe." The term "Metaverse" was first used in a piece of speculative fiction called Snow Crash by Neal Stephenson in 1992 [18].

The Metaverse, an evolving digital frontier, is rapidly transforming various sectors, daily activities, and the fabric of society itself, heralding immense possibilities for service enhancements. By seamlessly intertwining the virtual and physical worlds, the Metaverse allows users to craft and navigate their digital personas through avatars [19]. This paradigm shift is amplified by the integration of advanced technologies such as AI [20], ML, deep learning [21], DTs, IoT, edge computing, and cloud computing. Together, these innovations are driving the Metaverse's trajectory as a game-changing technological phenomenon.

4.2.2.1 Concept and Design of Metaverse

In the year 2021, Duan et al. presented a comprehensive three-tier architectural framework for the Metaverse [22] (Figure 4.2(a)). This seminal model articulates the Metaverse's structural organization, with a particular emphasis on the pivotal role played by the interaction layer in mediating the interconnection of the ecosystem and infrastructure components. Drawing upon Duan's original architectural proposal, the exposition distills the Metaverse's intricate stratification into a tripartite delineation, providing a coherent and structured overview of its seven constituent layers.

A. *Three-layer architecture:*
   1. *Infrastructure:* Essential to the virtual world, the infrastructure merges computation, communication, and blockchain technologies, grappling with high computational demands and decentralized data management.
   2. *Interaction:* Focusing on immersive experiences, interaction balances real-world data transfer, VR/AR interfaces, and DTs, emphasizing user-driven content creation and real-to-virtual object representation.
   3. *Ecosystem:* Powered by user-generated content [23, 24], the ecosystem intertwines decentralized finance, blockchain-based NFTs, and AI elements like NPCs, aspiring for a dynamic, parallel existence within the Metaverse.

B. *Seven-layer architecture:*

   The architecture of the Metaverse comprises seven primary layers, as illustrated in Figure 4.2(b). It is noteworthy that these seven layers can be regarded as a more detailed exposition of the three aforementioned layers within the architectural framework. The subsequent section expounds upon each of these layers in a comprehensive manner [25].

   1. *Experience:* The Metaverse transcends mere graphical dimensions, dematerializing the physical world into boundless virtual realms. From games like Fortnite to voice assistants like Alexa, it reshapes scarcities, making every concert seat the best. Users evolve from mere content consumers to creators, fueling an interconnected web of content and social immersion.

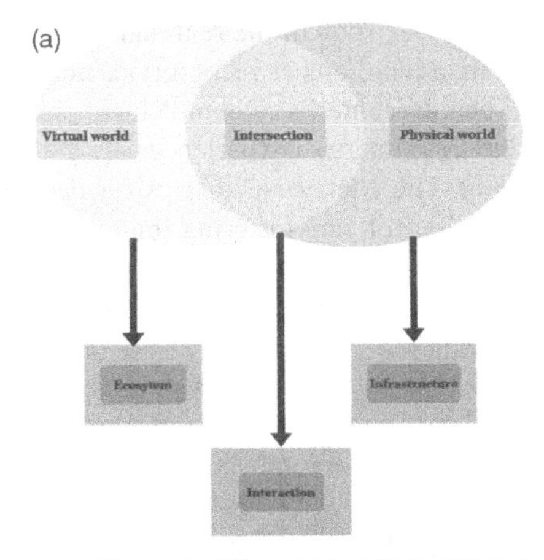

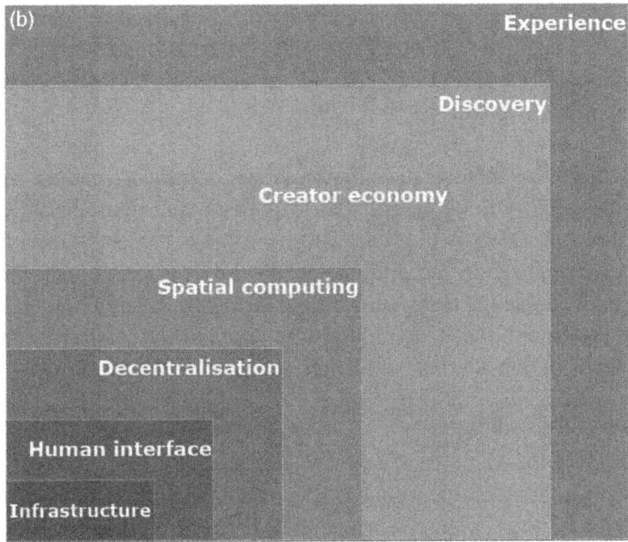

FIGURE 4.2 These figures illustrate the fundamental structure of the Metaverse. (a) Simplified three-layer architecture of the Metaverse and (b) comprehensive seven-layer architecture of the Metaverse.

2. *Discovery:* Discovery within the Metaverse is either active (inbound) or marketing-driven (outbound). Key elements like community-driven content and real-time presence become powerful discovery tools, with real-time interactions shifting focus from traditional social networking to dynamic social engagement.
3. *Creator economy:* An expanding universe of creators, aided by evolving tools, shape Metaverse experiences. Progressing from rudimentary creation stages, today's platforms, such as Roblox, provide holistic, integrated solutions. The Metaverse champions decentralized open platforms empowering creators.

4. *Spatial computing:* A fusion of the real and virtual, spatial computing offers 3D visualizations, object recognition, and real-time data integrations. Next-gen user interfaces complement concurrent information streams, enhancing immersive experiences.
5. *Decentralization:* The Metaverse thrives on decentralization, avoiding monopolistic control, and fostering innovation. From foundational technologies like DNS to the revolutionary blockchain, decentralization empowers user sovereignty, fluid markets, and scalable decentralized applications.
6. *Human interface:* Device evolution pushes us closer to a cyborg reality, from smartphones to VR like Oculus Quest. The future promises smart glasses, 3D-printed wearables, and potentially consumer neural interfaces, further blurring the lines between humans and machines.
7. *Infrastructure:* Supporting the Metaverse's expansive vision, the infrastructure layer boosts device capabilities, optimizes connectivity, and delivers seamless content. Innovations like 5G and beyond, advanced semiconductors, and compact power solutions form the backbone of this immersive world.

In the realm of the Metaverse, a dynamic and infinitely scalable virtual world unfolds, emphasizing both its capacity for expansion and interoperability. The digital counterparts, or DTs, of physical-world buildings, objects, and environments can be constructed through the application of 3D reconstruction techniques [26]. These techniques enable the translation of real-world entities into the virtual domain. However, the creation of these 3D models often requires specialized knowledge and experience, posing a considerable challenge for amateurs to replicate.

4.2.2.2 General Challenges

The construction of the Metaverse, a dynamic merger of virtual and real-world experiences, presents a range of open issues. These challenges span the fields of technology, ethics, privacy, and societal influence [27]. In this section, we go into key topics, explore the subtleties of interactivity, computing needs, ethical considerations, privacy precautions, and addressing the potential ramifications of prolonged digital immersion. Additionally, we investigate the growing landscape of virtual economies, social acceptability, security, and privacy, all while underlining the necessity for a fundamental framework that safeguards the rights of all users.

a. *Interaction challenges:*
    - Interaction devices should be lightweight, user-friendly, wearable, and portable.
    - The user experience should allow complete immersion, minimizing technological distractions.
    - Current technologies like somatosensory technology and extended reality (XR) (encompassing VR, AR, mixed reality) are not entirely transparent or lightweight and remain cost-prohibitive.

b. *Computation dynamics:*

- Metaverse requires heightened computational strength, involving data calculation, storage, and transmission.
- Rise of cloud technologies like cloud storage, computing, and rendering necessitate powerful device performance and robust server resilience.

c. *Ethical considerations:*
   - New identities and social structures in the Metaverse lead to complex societal interactions, needing clear ethical guidelines.
   - Issues include information integrity, fostering a positive atmosphere, and intellectual property rights.
   - Possible future advancements allowing manipulation of consciousness bring up significant ethical concerns.
   - Ensuring Metaverse regulations keep pace with technological advancements is imperative.

d. *Metaverse privacy and data protection:*
   - Direct linkage between real identities and Metaverse avatars intensifies data privacy concerns.
   - Given the intricate intertwining of the virtual and real worlds, protective measures are critical.

e. *Health implications: cyber-syndrome:*
   - Excessive internet use can lead to physical, social, and mental disorders known as cyber-syndrome.
   - The deep immersion in the Metaverse could exacerbate this issue due to the blurring lines between reality and the virtual world.

f. *Economic and social aspects:*
   - Reliability and adaptability of cryptocurrencies in the Metaverse.
   - Interdependence of virtual and real economies, urging a comprehensive perspective on consumption behaviors and economic activities in both realms.
   - Privacy threats, user diversity, fairness, and addiction can influence the Metaverse's sustainability.
   - Effective tools and technologies to combat cybercrime and misconduct are essential for fostering a safe environment.

g. *Security measures:*
   - Seamless authentication methods, moving away from traditional passwords, may involve alternative modalities such as biometric techniques.
   - Ensuring high levels of security, accurate detection, and user-friendly systems remains crucial.
   - Accumulated user interaction records pose potential long-term privacy risks.
   - Simplified, yet effective consent mechanisms, along with privacy-preserving ML, are required for dynamic user experiences.

h. *Trustworthiness and responsibility:*
   - The evolving nature of personal data in the Metaverse requires dynamic definitions that align with innovation.
   - As the Metaverse develops, it should prioritize minority and vulnerable community rights to prevent potential real-world repercussions.

4.3 DT-POWERED METAVERSE

The Metaverse offers a deeply immersive experience, hinging on a multitude of operations such as data collection, transmission, manipulation, and generation. This intricate web of interaction spans a range of devices from mobile phones and cameras to advanced helmets and edge nodes. At its core, the Metaverse is a product of cutting-edge technologies: next-generation networking, advanced communications, ML, AI, and DTs.

DTs, in particular, play a crucial role in overseeing the multifaceted operations of the Metaverse ecosystem. In essence, to effectively simulate and monitor the Metaverse, each operation, concept, and entity must have its own DT. These DTs, in tandem, offer a detailed and precise analysis of the Metaverse's functionalities. For instance, to ensure optimal service delivery, the DT of the Metaverse must seamlessly interact with the DT of the wireless network, which encapsulates the specifics of various network operations and infrastructure. Network performance significantly determines service quality.

Moreover, by forecasting and mapping the current and anticipated requests from Metaverse users onto the DT, there's an opportunity to better allocate network resources. This not only boosts the quality of service (QoS) but also enhances the system's autonomy.

One notable application where the Metaverse hopes to reap major benefits is in the field of DTs. The integration of DT with Metaverse services offers the ability to continually monitor user engagements and requisitions. Such integrative services would correctly account for the varied array of devices, basic concepts, and operational modes intrinsic to this domain, with network resources being of paramount consequence. Conversely, the formulation of DT for tangible entities can pave the way for the implementation of a real-world Metaverse, thereby boosting the authenticity and verisimilitude of the immersive experience.

In the intricacies of such a multifaceted ecosystem, the devised platform confronts a myriad of challenges to fulfill the stipulated requirements, ensuring seamless and secure interoperability across its diverse components. The constituent modules of the DT-enabled platform are strategically integrated within a distributed architectural framework. Within this framework, a plethora of edge nodes collaboratively function to ascertain the zenith of both quality of experience (QoE) and QoS. This collaborative regime is sustained by distinct system constituents and is communicated over a cutting-edge infrastructure, specifically the sixth-generation (6G) technology. It's imperative to note that both the DT and XR modules operate synergistically over these edge nodes. Furthermore, the governance of such modules predominantly resides at the edge. The requisition and assimilation of data from the tangible realm necessitates the utilization of 6G communication protocols to achieve superior service and experiential standards within Metaverse applications.

Aloqaily et al. have proposed a comprehensive layered platform architecture that serves as a significant reference point in this study [28]. Figure 4.3 delineates a comprehensive schema of the proposed DT-6G Metaverse platform. This avant-garde platform is architecturally segmented into various stratified layers and components: (i) IoT layer, (ii) DT layer, (iii) Metaverse managerial module, (iv) XR layer, and

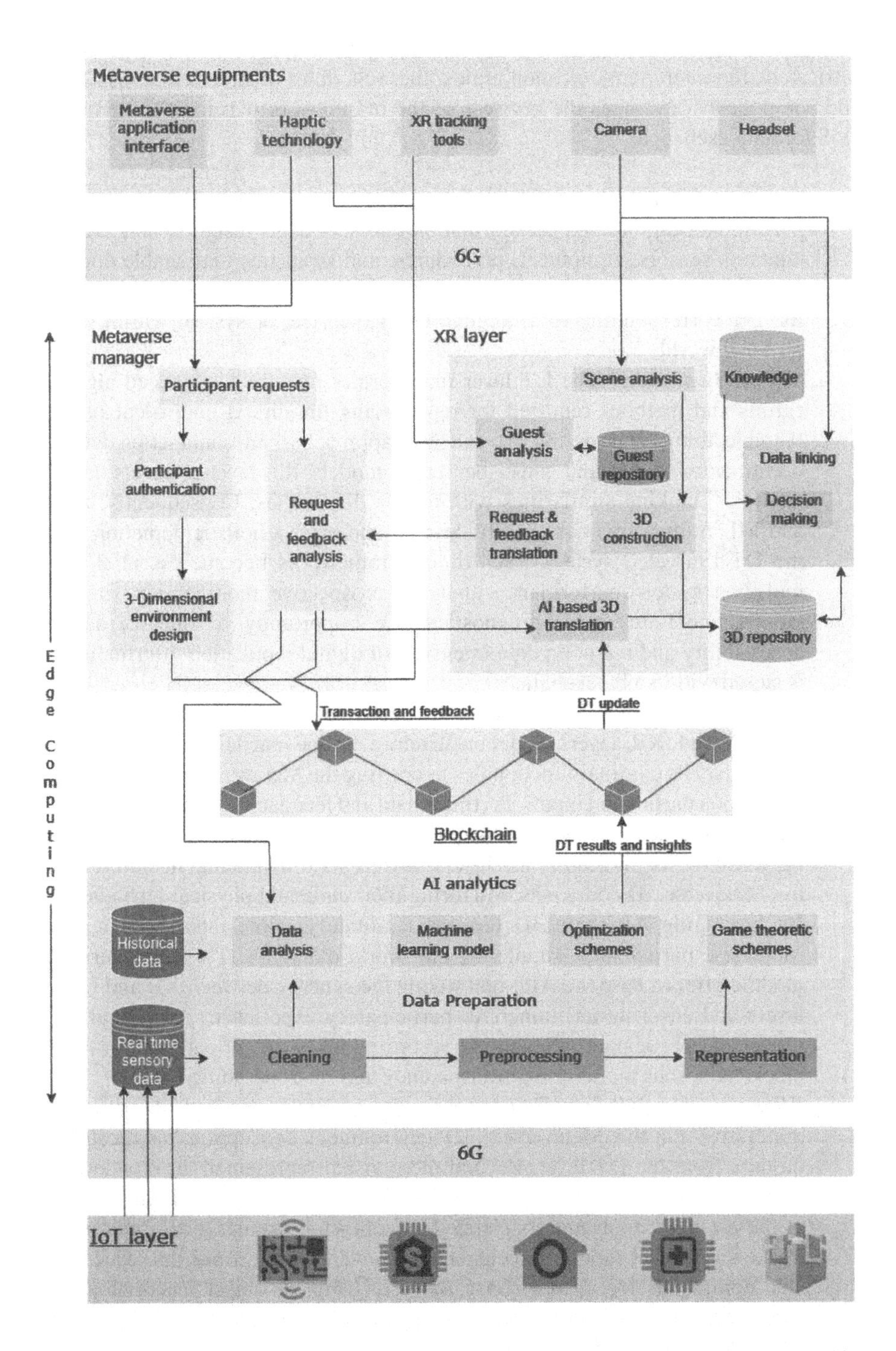

FIGURE 4.3 An extensive overview of the advanced Metaverse framework empowered by Digital Twin technology.

(v) blockchain. Each distinct layer is dedicated to a specific set of operations. Intricately, the intercommunication amidst these modules, distributed across layers and components, leverages the prowess of the 6G network, ensuring the requisite QoS for data exchanges.

i. *Internet of Things layer:* The IoT layer offers the basic infrastructure to permit data capture and transformation into actionable insights. This layer contains sensors, equipment, procedures, and strategies that enable comprehensive data collection. Such data serve as the bedrock for developing the DT corresponding to a tangible entity, service, or system within the physical world.
ii. *Digital Twin layer:* The DT layer incorporates numerous advanced algorithms and methods required for reproducing the digital equivalent of a tangible thing. Harnessing AI and ML approaches, in conjunction with optimization tactics and game-theoretical models, this layer conducts data analytics, making similarities with earlier discoveries. Consequently, AI and ML frameworks continually develop and offer a holistic depiction of the DT's targeted system. Such digital frameworks become essential in real-time system performance analysis, prospective modification simulations, and future event prognostications. Importantly, to enhance data accessibility and maintain data integrity, all digital replication information is stored within a blockchain.
iii. *Metaverse manager:* The Metaverse manager serves as the nexus between the DT and XR layers and concurrently bridge participants with the Metaverse. Its cardinal function lies in crafting the Metaverse terrain predicated upon participant inputs. Pertinent data and feedback are procured from the DT layer, conveyed via the blockchain, and subsequently transmitted to the XR layer for 3D rendition. The DT models are instrumental in architecting Metaverse 3D constructs, mirroring their authentic physical attributes. By leveraging AI, these 3D replicas facilitate genuine interaction with Metaverse participants, simulating real-world dynamics. This managerial module offers a bespoke API, optimizing the synergy between XR and DT layers and ensuring an immersive participatory experience. Additionally, it oversees participant interactions, recording all pertinent data within the blockchain, thus maintaining transparency and interoperability.
iv. *XR layer:* The XR layer houses the suite of operations and algorithms imperative for the Metaverse's 3D environment conception, predicated on data from the DT layer. Beyond mere visual representation, it endows users with an interactive experience imbued with realism. Utilizing state-of-the-art haptic tools and specialized sensors – for instance, ocular motion trackers – the XR modules decipher and convert user gestures into executable commands for virtual constructs. The tangible object reactions are sourced from the DT data encapsulated within the blockchain.
v. *Blockchain:* The verisimilitude of the Metaverse hinges on the meticulous extraction and amalgamation of voluminous data to construct DT replicas of real-world entities. Here, data accessibility and veracity emerge as pivotal

components in ensuring an unparalleled QoE. The integrated blockchain proffers myriad benefits for the DT-driven Metaverse (Figure 4.4). Not only does it vouchsafe the authenticity of each DT vis-à-vis their tangible counterparts, thereby validating data accuracy, its decentralized nature enables expeditious and transparent data access, negating third-party intervention. This facilitates the facile creation of a 3D milieu based on myriad DTs, allowing global participants to converge within a singular virtual realm. All interactions and pertinent Metaverse service data are cataloged within the blockchain, ensuring transparent model adjustments.

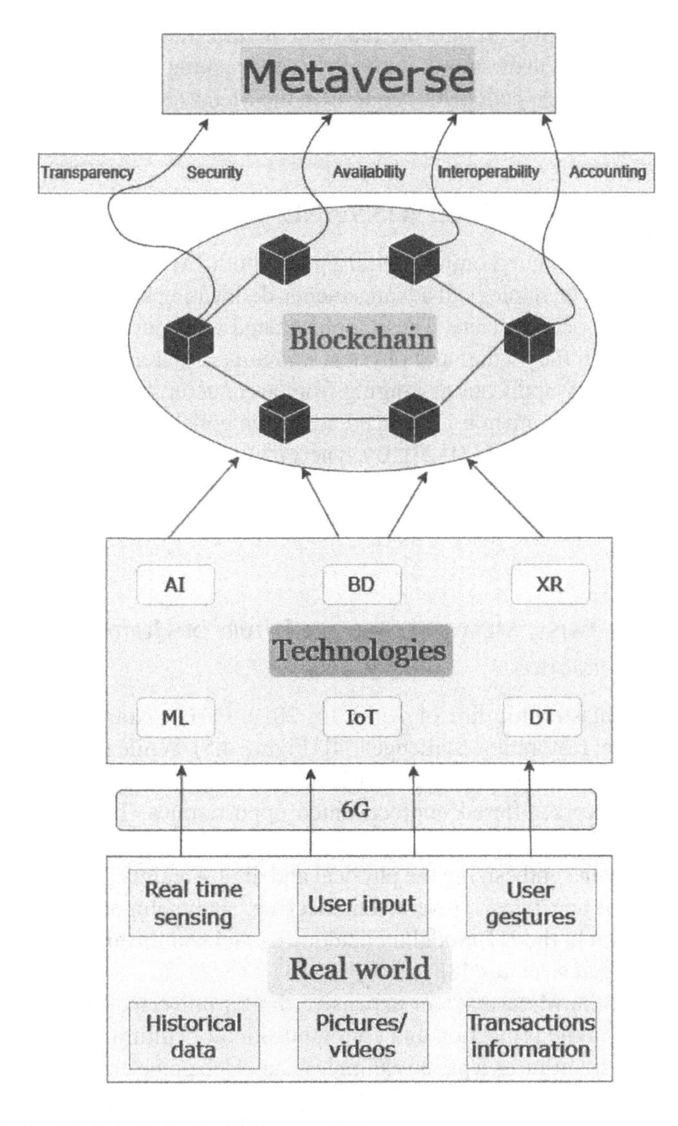

FIGURE 4.4 Blockchain in the DT-powered Metaverse.

4.3.1 Operation in the Real World

The implementation of a comprehensive real-world Metaverse is done through an integration of many technologies, especially DTs, XR, 6G, and blockchain. Initially, real-world data are captured and promptly sent via the 6G network to strategically selected edge nodes, factoring in their capability, workload, and user proximity [29]. Subsequent to data preprocessing at these nodes, AI and ML algorithms within the DT layer construct or refine the DT, with major alterations being authenticated and appended to a blockchain. As users declare the intent to enter the Metaverse, the Metaverse manager evaluates their requests, orchestrating the building of a corresponding 3D environment by extracting DT data from the blockchain. The XR layer, operating as the linchpin, permits the real-time manifestation and continual update of this environment. Furthermore, with advanced tracking techniques, users' interactions are decoded into actionable requests in the Metaverse, with AI-assisted logic providing dynamic and realistic object reactions within this digital realm.

4.4 IMPACT OF DIGITAL TWINS AND METAVERSE

In the context of our burgeoning digital epoch, both DTs and the Metaverse are emerging as salient technological advancements, delineating new horizons in myriad sectors of contemporary society. These sophisticated tools portend an era where the dichotomy between the virtual and physical domains is increasingly blurred, yielding multidisciplinary applications ranging from architectural heritage rehabilitation [30] and advanced urbanism to enhanced academic collaborations and avant-garde modes of digital engagement [31–33]. By synergizing precision with immersion, they are set to recalibrate our methodological frameworks, challenging established paradigms and engendering novel discourses in both technological and socio-cultural domains.

4.4.1 Digital Twins, Metaverse, and the Future of Heritage Reconstruction

Following the catastrophic fire of April 15, 2019, Paris's Notre-Dame Cathedral faced a significant restoration challenge [34] (Figure 4.5). While traditional methods provided insights, the convergence of digital solutions, primarily through the use of the DT framework, offered unprecedented opportunities. This technology was instrumental in the intricate reconstruction of the cathedral's collapsed transverse arch, particularly in synthesizing the physical and digital realms. Through integrated facets of physical anastylosis, reverse engineering, spatiotemporal annotation, and operational research, the DT model meticulously emulated the architectural intricacies of the damaged structure [8].

Furthermore, the Metaverse, an expansive digital universe, has been emerging as a pivotal tool in the protection and communication of cultural heritage. Take the Notre-Dame's restoration as a prime example. The video game company Ubisoft had recreated the cathedral as an intricate 3D model in their game, "Assassin's Creed Unity." This digital model, although initially designed for entertainment purposes,

FIGURE 4.5 Photo of Notre-Dame Cathedral in Paris on April 15, 2019, taken from Quai de Montebello (credit: Wandrille de Préville). Source: [35].

has become an invaluable resource for the monument's reconstruction endeavors. Such serendipitous digital preservations emphasize the Metaverse's potential in safeguarding fragile cultural relics against human-induced damages or unforeseeable natural calamities.

Looking eastward, similar efforts are discernible. In China, collaborations between national architects and platforms like Cthuwork [36] have resulted in the digital reincarnation of monumental cultural relics. Buildings such as the Forbidden City and iconic artworks like "Qingming Shanghe Tu" have been revitalized as 3D voxel models in the popular game Minecraft. These digital reconstructions, accessible from anywhere globally, serve dual purposes. They act as immersive gateways to historical wonders for the layperson and invaluable references for experts engaged in actual restoration projects.

In essence, the synergy between DTs and the Metaverse not only aids in meticulous reconstructions but also envisions a future where heritage preservation transcends traditional constraints, fostering a seamless blend of the tangible and virtual realms.

4.4.2 Digital Twin and Metaverse in Healthcare

The confluence of DTs and the Metaverse in healthcare, buttressed by an array of technologies, promises unparalleled patient-centric care paradigms [37–39]. As major organizations invest heavily in the Metaverse platform, the healthcare sector stands on the brink of a digital revolution, ensuring precision and personalized care delivery.

4.4.2.1 Digital Twins

The integration of DTs in healthcare signifies a transformative approach that can potentially usher in individualized patient care. Derived from a comprehensive review of medical literature, the significance of DTs emerges as a vital asset for enhancing the already well-established personalized treatment methodologies, especially in cancer care [40].

A healthcare DT represents a virtual replica of an individual, amalgamating a lifetime of personal health data through AI models. This approach aids in predicting health trajectories, making it pivotal for predicting outcomes of procedures. With the integration of the Medical Internet of Things, the growth trajectory of healthcare DTs is robust, forecasting a market valuation of approximately USD 75.44 billion within the next decade [41, 42].

ML has already displayed promise in enhancing DT capabilities. Notable studies, for instance, utilized artificial neural networks for risk assessment pertaining to deep vein thrombosis and pulmonary embolism based on a host of factors. These networks, once optimized, can negate the need for certain medical procedures by predicting conditions with higher accuracy. Moreover, another investigation formulated a decision model, using ML for CT imaging of pulmonary embolism patients [43], ensuring efficient data use. However, the complexity and sensitivity of medical data introduce challenges for DT implementations, demanding ease of use and adaptability for frontline health professionals, including nurses and doctors.

4.4.2.2 Metaverse

In their 2023 scholarly publication, Turab and Jamil provided an exhaustive analysis of the requisite technologies and the application of the Metaverse in healthcare [44]. What follows is a concise synthesis of their comprehensive study. The Metaverse, with its multifaceted applications in healthcare, represents a digital frontier poised to reshape healthcare methodologies. To navigate this digital realm, a plethora of enabling technologies are vital, such as:

i. *Computer vision (CV):* Fundamental for in-situ disease diagnosis and intricate medical imaging.
ii. *IoT:* Provides crucial surgical assistance, ensuring timely alerts and information delivery.
iii. *Human–computer interface:* Optimizes remote medical assistance and service delivery.
iv. *AI and quantum computing:* Key for deriving insights, making decisions, and ensuring high computational speed alongside quantum-resistant security for medical applications.
v. *Blockchain, Big Data, and 5G:* Ensures data integrity, advances healthcare data management, and delivers unmatched immersive communication experiences.
vi. *Extended reality and 3D modeling:* Provides a virtual realm for training, consultation, and intricate anatomical visualization.
vii. *Edge computing:* Amplifies data transfer rates and analytical capabilities.

Another cardinal component of the Metaverse is the VR which encompasses immersive, interactive 3D environments. Incorporating VR and other Metaverse technologies, especially when intertwined with DTs, could herald a paradigm shift, particularly in the diagnosis and treatment of conditions like cancer.

VR classifications include:

i. *Non-immersive*: Typically, 3D simulations on computer screens, like games or interactive design platforms.
ii. *Semi-immersive*: Limited VR experiences, with flight simulators being paradigmatic examples.
iii. *Fully immersive*: Represents the zenith of VR immersion, with users being wholly submerged in virtual environments using specialized gear.

Thus, the integration of DTs and the Metaverse into healthcare signals a transformative shift, leveraging cutting-edge technologies to offer more precise, individualized care and harnessing the vast potential of virtual realms for therapeutic and diagnostic advancements.

4.4.3 Additional Applications of Metaverse and Digital Twins

In addition to the primary focus areas, there are myriad applications of the Metaverse and DTs that, while not delved into deeply within this chapter, merit acknowledgment due to their significance in contemporary technology and business landscapes.

A. Digital Twins applications: DTs are increasingly refined by developers and researchers, given their vast application potential across various domains:

i. *Industry and Science*: Enhancing instrument predictions, risk mitigation, and understanding product reactions in diverse environments, ensuring minimal tangible risks.
ii. *Mobility*: Optimizing smart vehicles for better performance and safety.
iii. *IoT & Industrial IoT (IIoT):* Enhancing system output and operational efficiency.
iv. *Industry 4.0*: Advancing manufacturing processes and automation [45, 46].

Practical implementations of DT encompass a spectrum from accelerating product development cycles in tech and science sectors, crafting sophisticated retail market strategies, predicting climatic patterns, to extensive applications in various business spheres. It's noteworthy that these examples represent just a fraction of DT's vast potential.

B. Metaverse applications: Beyond the discussed domains, the Metaverse has broadened its reach into diverse areas:

i. *Military*: Utilization of Tactical Augmented Reality (TAR) for advanced combat and strategic planning [47].
ii. *Real estate*: Enhanced property visualization through VR.
iii. *Manufacturing*: Streamlining processes and virtual prototyping.

iv. *Education*: Interactive learning experiences using VR headsets [48].
v. *Travel & shopping*: Virtual tourism and immersive e-commerce experiences [49, 50].
vi. *Virtual collaboration*: Facilitating virtual meetings and conferences.

Furthermore, in the entertainment and business sectors, film producers' premiere trailers within the Metaverse, fashion brands host virtual runway shows, e-commerce platforms incorporate real-sense feedback, corporations coordinate virtual commercial meetings, and game developers showcase their innovative Metaverse creations.

4.5 CONCLUSION

In the dynamic landscape of technological evolution, the harmonious convergence of DT and the Metaverse emerges as a potent force, capable of reshaping the contours of multiple sectors. This chapter has embarked on an exhaustive journey, traversing their origins, current trajectories, and aspirational future implications, shedding light on their potential to revolutionize both traditional and contemporary systems.

Historical evolutions, such as DT's embryonic stages during NASA's space missions and the burgeoning rise of Metaversal platforms, emphasize not just their adaptability, but also their escalating significance in the current technological milieu. Such advancements underscore their potential to transcend boundaries, merging the tangible and intangible, as exemplified by endeavors like the meticulous digital resurrection of cultural icons like Paris's Notre-Dame Cathedral. These endeavors demonstrate a transformative approach to preserving global heritage, with DT and the Metaverse acting as pivotal tools.

The synergy of DT and the Metaverse holds promise beyond heritage conservation. In healthcare, for instance, this alliance paves the path for innovative patient-centric paradigms that lean on precision, predictability, and real-time responsiveness. These integrative methodologies could revolutionize treatment trajectories, forecasting health conditions, and tailoring interventions with unprecedented accuracy.

The complexities and subtleties associated with DTs and the Metaverse are undeniably vast, and their continuous evolution demands acute attention. Their confluence hints at a foreseeable future where the delineation between the physical and virtual realms dissipates, crafting an intertwined reality. As we venture deeper into this novel digital frontier, it is paramount for stakeholders, ranging from technologists to policymakers, to be at the helm, guiding this technological amalgamation toward outcomes that are not only innovative but also equitable and sustainable.

To encapsulate, this exploration unveils a tantalizing glimpse into our imminent digital epoch. Positioned at this transformative juncture, it remains our shared duty to shepherd this symbiotic relationship between DT and the Metaverse, ensuring a future that resonates with innovation, inclusivity, and unparalleled possibilities.

REFERENCES

1. P. A. Cola, K. Lyytinen and S. A. Nartker, Voices of Practitioner Scholars in Management: The History and Impact of the Doctor of Management Programs at Case Western Reserve University. Wilmington, OH: Weatherhead School of Management, Case Western Reserve University by Orange Frazer Press, 2020.
2. N. Crespi, A. T. Drobot and R. Minerva, The Digital Twin. Cham: Springer, 2023.
3. R. Minerva and N. Crespi, "Digital Twins: Properties, software frameworks, and application scenarios," IT Professional, vol. 23, no. 1, pp. 51–55, 2021. doi: 10.1109/MITP.2020.2982896.
4. R. Minerva, F. M. Awan and N. Crespi, "Exploiting Digital Twins as enablers for synthetic sensing," IEEE Internet Computing, vol. 26, no. 5, pp. 61–67, 2022. doi: 10.1109/mic.2021.3051674.
5. T. Deng, K. Zhang and Z.-J. Shen, "A systematic review of a Digital Twin City: A new pattern of urban governance toward smart cities," Journal of Management Science and Engineering, vol. 6, no. 2, pp. 125–134, 2021. doi: 10.1016/j.jmse.2021.03.003.
6. S. B. Far and A. I. Rad, "Applying Digital Twins in Metaverse: User interface, security and privacy challenges," Journal of Metaverse, vol. 2, no. 1, pp. 8–15, 2022. doi: 10.48550/arXiv.2204.11343.
7. F. Tao, H. Zhang, A. Liu and A. Y. C. Nee, "Digital Twin in industry: State-of-the-art," IEEE Transactions on Industrial Informatics, vol. 15, no. 4, pp. 2405–2415, 2019. doi: 10.1109/TII.2018.2873186.
8. Borole, Yogini, Borkar, Pradnya, Raut, Roshani, Balpande, Vijaya Parag and Chatterjee, Prasenjit. *Digital Twins: Internet of Things, Machine Learning, and Smart Manufacturing*, Berlin, Boston: De Gruyter, 2023. https://doi.org/10.1515/9783110778861.
9. K. Dröder, P. Bobka, T. Germann, F. Gabriel and F. Dietrich, "A machine learning-enhanced Digital Twin approach for human–robot collaboration," Procedia CIRP, vol. 76, pp. 187–192, 2018, doi: 10.1016/j.procir.2018.02.010.
10. M. Xia, et al., "Intelligent fault diagnosis of machinery using Digital Twin-assisted deep transfer learning," Reliability Engineering and System Safety, vol. 215, p. 107938, 2021. doi: 10.1016/j.ress.2021.107938.
11. Q. Qi, D. Zhao, T. W. Liao, and F. Tao, "Modeling of cyber-physical systems and digital twin based on edge computing, fog computing and cloud computing towards smart manufacturing," in *ASME 2018 13th International Manufacturing Science and Engineering Conference*, vol. 1, pp. 1–7, 2018. doi: 10.1115/msec2018-6435.
12. V. Souza, R. Cruz, W. Silva, S. Lins and V. Lucena, "A Digital Twin architecture based on the Industrial Internet of Things technologies," in 2019 IEEE International Conference on Consumer Electronics (ICCE), 2019. doi: 10.1109/icce.2019.8662081.
13. J. Leng, et al., "Digital Twins-driven manufacturing cyber-physical system for parallel controlling of smart workshop," Journal of Ambient Intelligence and Humanized Computing, vol. 10, no. 3, pp. 1155–1166, 2018. doi: 10.1007/s12652-018-0881-5.
14. S. Aheleroff, X. Xu, R. Y. Zhong and Y. Lu, "Digital Twins as a service (DTaaS) in industry 4.0: An architecture reference model," Advanced Engineering Informatics, vol. 47, p. 101225, 2021. doi: 10.1016/j.aei.2020.101225.
15. M. W. Grieves, "Digital Twins: Past, present, and future," The Digital Twin, pp. 97–121, 2023. doi: 10.1007/978-3-031-21343-4_4.
16. A. Fuller, Z. Fan, C. Day and C. Barlow, "Digital Twin: Enabling technologies, challenges and open research," IEEE Access, vol. 8, pp. 108952–108971, 2020. doi: 10.1109/access.2020.2998358.
17. R. Vishwakarma and A. K. Jain, "A survey of DDoS attacking techniques and defence mechanisms in the IOT network," Telecommunication Systems, vol. 73, no. 1, pp. 3–25, 2019. doi: 10.1007/s11235-019-00599-z.

18. J. Joshua, "Information bodies: Computational anxiety in Neal Stephenson's snow crash," Interdisciplinary Literary Studies, vol. 19, no. 1, pp. 17–47, 2017. doi: 10.5325/intelitestud.19.1.0017.
19. B. Carey, "Metaverse technologies, behavioral predictive analytics, and customer location tracking tools in blockchain-based virtual worlds," Review of Contemporary Philosophy, vol. 21, p. 188, 2022. doi: 10.22381/rcp21202212
20. T. Huynh-The, et al., "Artificial intelligence for the Metaverse: A survey," Engineering Applications of Artificial Intelligence, vol. 117, p. 105581, 2023. doi: 10.1016/j.engappai.2022.105581.
21. S. Jamil, et al., "Deep learning and computer vision-based a novel framework for Himalayan Bear, Marco Polo Sheep and snow leopard detection," in 2020 International Conference on Information Science and Communication Technology (ICISCT), 2020. doi: 10.1109/icisct49550.2020.9080021.
22. H. Duan, et al., "Metaverse for social good," in Proceedings of the 29th ACM International Conference on Multimedia, pp. 153–161, 2021. doi: 10.1145/3474085.3479238.
23. P. Ludlow and M. Wallace, The Second Life Herald: The Virtual Tabloid That Witnessed the Dawn of the Metaverse. Cambridge, MA: MIT Press, 2009.
24. J. Krumm, N. Davies and C. Narayanaswami, "User-generated content," IEEE Pervasive Computing, vol. 7, no. 4, pp. 10–11, 2008. doi: 10.1109/mprv.2008.85.
25. T. Min, H. Wang, Y. Guo and W. Cai, "Blockchain games: A survey," in 2019 IEEE Conference on Games (CoG), 2019. doi: 10.1109/cig.2019.8848111.
26. Z. Ma and S. Liu, "A review of 3D reconstruction techniques in civil engineering and their applications," Advanced Engineering Informatics, vol. 37, pp. 163–174, 2018. doi: 10.1016/j.aei.2018.05.005.
27. A. S. Bale, et al., "A comprehensive study on Metaverse and its impacts on humans," Advances in Human–Computer Interaction, vol. 2022, pp. 1–11, 2022. doi: 10.1155/2022/3247060.
28. M. Aloqaily, O. Bouachir, F. Karray, I. Al Ridhawi and A. E. Saddik, "Integrating Digital Twin and advanced intelligent technologies to realize the Metaverse," IEEE Consumer Electronics Magazine, vol. 12, no. 6, pp. 47–55, 2023. doi: 10.1109/mce.2022.3212570.
29. M. Aloqaily, O. Bouachir and I. A. Ridhawi, "Blockchain and FL-based network resource management for interactive immersive services," in 2021 IEEE Global Communications Conference (GLOBECOM), 2021. doi: 10.1109/globecom46510.2021.9685091.
30. X. Zhang, et al., "Metaverse for cultural heritages," Electronics, vol. 11, no. 22, p. 3730, 2022. doi: 10.3390/electronics11223730.
31. A. Novák, T. Klieštik and G. Lăzăroiu, "Live shopping in the Metaverse: Visual and spatial analytics, cognitive artificial intelligence techniques and algorithms, and immersive digital simulations," Linguistic and Philosophical Investigations, vol. 21, p. 187, 2022. doi: 10.22381/lpi21202212.
32. L. Rydell, "Predictive algorithms, data visualization tools, and artificial neural networks in the retail Metaverse," Linguistic and Philosophical Investigations, vol. 21, p. 25, 2022. doi: 10.22381/lpi2120222.
33. K. Valášková, V. Machová and E. Lewis, "Virtual marketplace dynamics data, spatial analytics, and customer engagement tools in a real-time interoperable decentralized Metaverse," Linguistic and Philosophical Investigations, vol. 21, p. 105, 2022. doi: 10.22381/lpi2120227.
34. W. K. Tannous, "The fire of Notre Dame: Economic lessons learned," WIT Transactions on The Built Environment, 2019. doi: 10.2495/dman190051.
35. W. de Préville, Français: Photo de Notre-Dame de paris le 15 avril 2019 à 19h17 prise du quai de montebello, Apr. 15, 2019. Wikipedia (accessed: Sep. 25, 2023). https://commons.wikimedia.org/wiki/File:NotreDame20190415QuaideMontebello_(cropped).jpg

36. H. Duan, et al., "Metaverse for social good," in Proceedings of the 29th ACM International Conference on Multimedia, p. 155, 2021. doi: 10.1145/3474085.3479238.
37. A. Bandyopadhyay, A. Sarkar, S. Swain, D. Banik, A. E. Hassanien, S. Mallik, A. Li and H. Qin, "A game-theoretic approach for rendering immersive experiences in the Metaverse", Mathematics, vol. 11, no. 6, p. 1286, 2023.
38. A. Sihna, H. Raj, R. Das, A. Bandyopadhyay, S. Swain and S. Chakrborty, "Medical education system based on Metaverse platform: A game theoretic approach", in IEEE 4th International Conference on Intelligent Engineering and Management (ICIEM 2023), 2023, pp. 1–6.
39. P. Gupta, K. Bhadani, A. Bandyopadhyay, D. Banik and S. Swain, "Impact of Metaverse in the near 'future'", in IEEE 4th International Conference on Intelligent Engineering and Management (ICIEM 2023), 2023, pp. 1–6.
40. O. Moztarzadeh, et al., "Metaverse and healthcare: Machine learning-enabled Digital Twins of cancer," Bioengineering, vol. 10, no. 4, p. 455, 2023. doi: 10.3390/bioengineering10040455.
41. G. Wang, et al., "Development of Metaverse for intelligent healthcare," Nature Machine Intelligence, vol. 4, no. 11, pp. 922–929, 2022. doi: 10.1038/s42256-022-00549-6.
42. W. Zhou, Y. Jia, A. Peng, Y. Zhang and P. Liu, "The effect of IoT new features on security and privacy: New threats, existing solutions, and challenges yet to be solved," IEEE Internet of Things Journal, vol. 6, no. 2, pp. 1606–1616, 2019. doi: 10.1109/jiot.2018.2847733.
43. I. Banerjee, et al., "Development and performance of the pulmonary embolism result forecast model (perform) for computed tomography clinical decision support," JAMA Network Open, vol. 2, no. 8, 2019. doi: 10.1001/jamanetworkopen.2019.8719.
44. M. Turab and S. Jamil, "A comprehensive survey of Digital Twins in healthcare in the era of Metaverse," BioMedInformatics, vol. 3, no. 3, pp. 563–584, 2023. doi: 10.3390/biomedinformatics3030039.
45. F. Pires, A. Cachada, J. Barbosa, A. P. Moreira and P. Leitao, "Digital Twin in industry 4.0: Technologies, applications and challenges," in 2019 IEEE 17th International Conference on Industrial Informatics (INDIN), 2019. doi: 10.1109/indin41052.2019.8972134.
46. J. Leng, et al., "Digital Twins-based smart manufacturing system design in industry 4.0: A review," Journal of Manufacturing Systems, vol. 60, pp. 119–137, 2021. doi: 10.1016/j.jmsy.2021.05.011.
47. K. Bunnag, "Alternate 'Realities': Military applications in the Metaverse era," DTAJ, vol. 5, no. 11, pp. 26–37, 2023.
48. P. Onu, A. Pradhan and C. Mbohwa, "Potential to use Metaverse for future teaching and learning," Education and Information Technologies, 2023. doi: 10.1007/s10639-023-12167-9.
49. T. Jenkins, "Immersive virtual shopping experiences in the retail Metaverse: Consumer-driven e-commerce, blockchain-based digital assets, and data visualization tools," Linguistic and Philosophical Investigations, vol. 21, p. 154, 2022. doi: 10.22381/lpi21202210.
50. S. Tsai, "Investigating Metaverse marketing for travel and tourism," Journal of Vacation Marketing, p. 135676672211457, 2022. doi: 10.1177/13567667221145715.

5 Metaverse for Development of Next-Generation Smart Cities

Sujata Swain
School of Computer Engineering, Kalinga Institute of Industrial Technology, Bhubaneswar, India

Anjan Bandyopadhyay
School of Computer Engineering, Kalinga Institute of Industrial Technology, Bhubaneswar, India

Rina Kumari
School of Computer Engineering, Kalinga Institute of Industrial Technology, Bhubaneswar, India

Roshan Chatei
School of Computer Engineering, Kalinga Institute of Industrial Technology, Bhubaneswar, India

Rudrashish Das
School of Computer Engineering, Kalinga Institute of Industrial Technology, Bhubaneswar, India

5.1 INTRODUCTION

Meta, formerly known as Facebook, has created a global computerized platform known as the Metaverse, which envisions a connected three-dimensional network of virtual realms known as the multiverse [1]. The Metaverse could provide residents of cities unsuitable for human existence with new creative opportunities. Recent technological and computing advancements have made this platform conceivable. However, the creation of this network raises ethical issues including privacy, data surveillance, human health and well-being, collective and cognitive echochambers, and geosurveillance.

People's day-to-day lives, jobs, and relationships have all been profoundly altered as a result of the widespread dissemination of COVID-19 across the globe. It is anticipated that the Metaverse will play an increasingly crucial role in postpandemic

 DOI: 10.1201/9781003449256-5

urban society as people explore new ways to connect and interact with one another distantly. The concept of what it means to live in a city is evolving as a result of the proliferation of activities that may be enjoyed by individuals without having to leave the confinement of their houses. The creation of the Metaverse laid the groundwork for this to become possible. This has the potential to diminish the impact of urbanization as well as lower the amount of land that is required [2].

The Metaverse has the ability to welcome a brand new era of economic opportunity and bring about a dramatic shift in the way that we go about our day-to-day job. The proliferation of remote work opportunities made possible by technological advancements has the potential to raise levels of productivity while simultaneously lowering the demand for conventional workplaces. As a direct consequence of the Metaverse, there have been expressions of worry regarding the prospects of employment as well as the continued existence of traditional businesses [3].

This research will look at the potential effects of the Metaverse, a virtual representation of data-driven smart cities, on city people's lifestyles, as well as the ethical issues that this transition raises. This chapter [4] will look at the most recent developments, studies, and trends in data-driven smart cities, urban informatics, and urban studies.

In order to accomplish that goal, the approach that will be utilized in this chapter will be one that is based around an exhaustive analysis of recent research and existing patterns. We will look at how these concerns could affect people's rights to privacy. The study that is given here is intended to enlighten policymakers about the risks that are posed by the Metaverse. This will allow them to develop policies that take into consideration the benefits of socially disruptive technology while also reducing the negative implications. Table 5.1 shows the key characteristics and implication of Metaverse in smart cities.

The Metaverse is a nascent area of inquiry that has the capacity to significantly transform our perceptions and interactions with the digital realm. The Metaverse possesses the capacity to enhance the verisimilitude of simulated surroundings. However, in order to ensure that the benefits of this technology can be realized while simultaneously mitigating any potential negative effects it may have, it is necessary to address the ethical concerns that are raised by this technology. This chapter includes a detailed examination of the ethical difficulties that are created by the Metaverse and offers recommendations for policymakers in order to ensure that this technology is developed in a way that benefits society. This will ensure that this technology is developed in a way that is beneficial to society.

5.2 THE METAVERSE AS A VIRTUAL FORM OF DATA-DRIVEN SMART CITIES

Smart cities that are data-driven employ advanced technologies like the Internet of Things, big data analytics, and machine learning to collect, analyze, and manage vast amounts of data pertaining to their inhabitants and the surrounding environment [1, 4]. The denizens of these urban centers can utilize these data to arrive at more informed choices for themselves and their localities. The data that have been obtained are utilized to enhance the operational efficiency of the city, improve the quality of care provided

TABLE 5.1
Key Characteristics and Implications of Incorporating the Metaverse in Smart Cities

Key characteristics	Implications
Data-driven	The Metaverse allows for vast amounts of data to be collected and analyzed, which can improve the efficiency and sustainability of smart cities. However, this also raises concerns about data privacy and security, as well as the potential for surveillance
Virtual	The Metaverse allows for virtual experiences that can supplement or replace physical experiences in smart cities, which can lead to greater accessibility and inclusivity. However, this also raises concerns about the potential for digital isolation and the erosion of face-to-face communication skills
Computational	The Metaverse uses algorithms and artificial intelligence to analyze data and provide personalized experiences, which can improve the quality of life in smart cities. However, this also raises concerns about the potential for bias and discrimination in decision-making
User-centered	The Metaverse is designed to cater to the needs and preferences of its users, which can lead to greater satisfaction and engagement in smart cities. However, this also raises concerns about the potential for addiction and mental health concerns
Networked	The Metaverse is a networked system that connects users from all over the world, which can lead to greater collaboration and innovation in smart cities. However, this also raises concerns about the potential for cybercrime and online harassment

to its inhabitants, and promote the overall prosperity of the city [4]. Figure 5.1 shows the usage of data-driven smart city applications according to smart city indicators.

5.2.1 The Potential Impact of the Metaverse on Physical Cities and Urban Development

The Metaverse's growing prominence is anticipated to have significant and enduring impacts on urban development [1], with potential positive or negative consequences contingent upon the outcomes of these effects [3, 5–7]. The Metaverse harbors the potential to enhance the efficacy and sustainability of urban regions in the long run. The utilization of virtual simulations that replicate authentic environments presents a promising opportunity to aid urban planners in analyzing diverse scenarios and models for the advancement of urban regions. The proposed approach could facilitate the process of identifying the most efficacious and enduring solutions prior to any tangible implementation. The utilization of this particular approach possesses the capability to alleviate the adverse impacts of urbanization on the natural surroundings, while concurrently enhancing the standard of living and general well-being of urban inhabitants [1, 8].

Nonetheless, it is crucial to take into account the possible adverse effects. A significant apprehension pertains to the potential of the Metaverse to intensify the centralization of authority and affluence in metropolitan regions, thereby aggravating preexisting

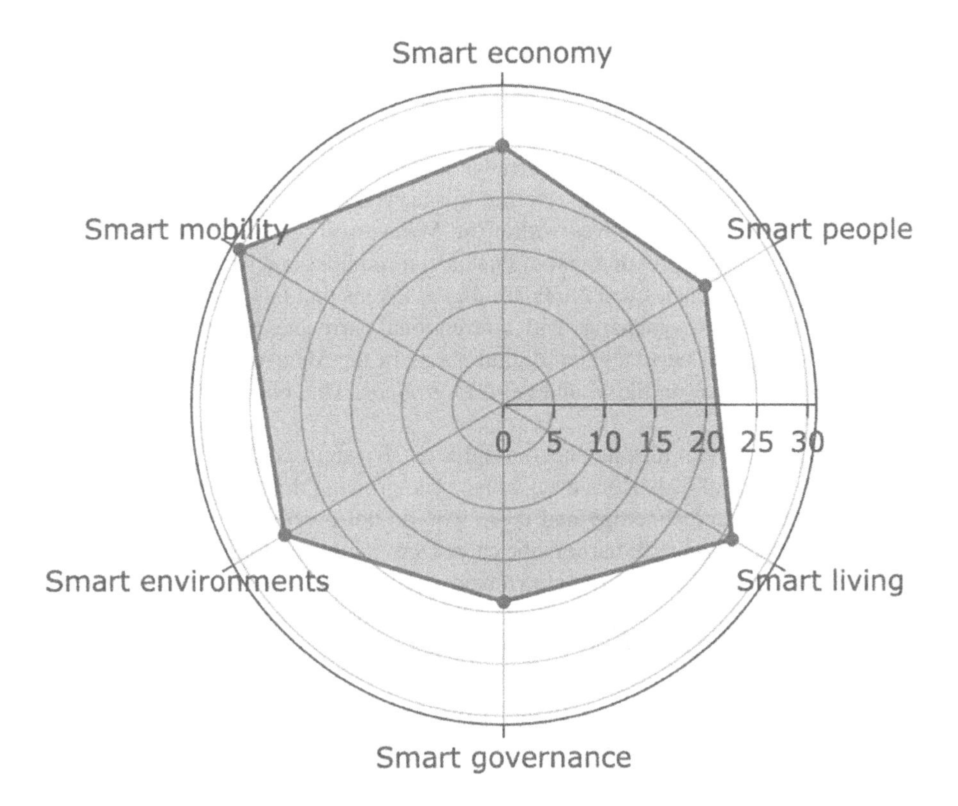

FIGURE 5.1 Usage of data-driven smart city applications according to smart city indicators.

societal disparities. In the event that entry to the Metaverse is restricted to individuals possessing the financial resources to procure it, this may engender a novel digital gap that bolsters preexisting societal and financial inequalities [1, 3, 5–7].

The growing dependence on data-driven models and algorithms within the Metaverse gives rise to significant ethical inquiries regarding the character of urban development. An illustrative query pertains to the determination of the individuals or entities responsible for selecting the data utilized in these models, as well as the methods employed to gather and evaluate such data. What ethical considerations arise from utilizing these models in the context of urban planning and development decision-making? What measures can be taken to guarantee the transparency and accountability of these models [1, 3, 5–7]?

The Metaverse's potential impact on physical cities and urban development is a complex and multifaceted issue that necessitates meticulous consideration and analysis. Although this technology presents certain advantages, it is crucial to acknowledge its possible adverse effects and establish guidelines and protocols to alleviate such risks. By using this approach, we may encourage the incorporation of ethical issues in the creation of the Metaverse, which will lead to the establishment of urban environments that are more sustainable, equitable, and welcoming for all parties concerned [1, 3, 5–7].

5.2.2 Virtual Currencies and Other Digital Assets in the Metaverse and Their Potential to Worsen Economic and Social Inequality

The emergence of the Metaverse presents novel prospects, such as the creation of alternative assets and virtual currencies, which exemplify alternative assets. Nonfungible tokens (NFTs), also referred to as cryptocurrencies, are being utilized to accomplish diverse objectives within the Metaverse. The objectives encompass trade, investment, and procurement of virtual commodities and services, as stated in Ref. [9]. Cryptocurrencies and NFTs are digital tokens that lack fungibility, rendering them unsuitable for exchange with conventional currencies or commodities. The utilization of virtual currency and digital assets in the Metaverse has encountered substantial resistance on moral and cultural grounds. This phenomenon of opposition is present.

One of the most significant concerns is that these resources can make the existing economic and social disparities even worse. The growing difference between those who have access to technology and those who do not is referred to as the "digital divide" [10]. The Metaverse harbors the capacity to facilitate global communication. However, the utilization of virtual currencies and digital assets may exacerbate the socioeconomic gap between individuals who possess the financial means to invest in them and those who do not. It is feasible that a system that is analogous to the one that exists in the actual world, in which power and wealth are concentrated around a small group of people, may arise in the virtual world [11].

In addition, the utilization of virtual currencies and digital assets gives rise to inquiries regarding regulatory measures and liability. Due to the absence of physical backing, nontraditional currencies and assets are frequently exempt from the same regulatory measures as their traditional counterparts [9]. The absence of regulatory measures may potentially result in fraudulent activities, money laundering, and other illicit behaviors that could have tangible ramifications [10]. Moreover, owing to its decentralized and distributed nature [12], regulating and enforcing regulations on the utilization of these assets within the Metaverse may pose a formidable challenge.

The utilization of virtual currencies and digital assets in the Metaverse raises ethical apprehensions regarding the possibility of exploitation and manipulation. The susceptibility of virtual currencies and digital assets to speculation, hype, and price volatility has been well documented and may result in the emergence of bubbles and subsequent crashes [9]. Individuals who lack a comprehensive comprehension of the workings of Metaverse assets may be at a heightened risk of being exploited and manipulated by proficient users or individuals who aim to capitalize on their lack of knowledge.

The utilization of virtual currencies and digital assets within the Metaverse elicits a number of ethical and societal considerations. The aforementioned assets possess the capacity to amplify preexisting economic and societal disparities and establish a digital marketplace that reflects the tangible economy [11]. Furthermore, the absence of proper oversight and responsibility may result in illicit conduct [10], whereas the instability and conjecture encompassing these commodities may result in abuse and maneuvering [9]. The formulation of policies and regulations that strike a balance between the advantages and potential hazards of these assets is imperative to guarantee a just and impartial Metaverse for all.

5.2.3 The Metaverse's Ability to Change Power Structures and Social and Political Order, and the Ethical Issues Involved

It is possible that as the Metaverse continues to advance, it will force existing power structures to shift, paving the way for the development of new social and political systems [3]. It is hoped that the Metaverse would facilitate decentralization and spark innovative community-based forms of collaboration.

The potential of the Metaverse to generate innovative social and economic structures and facilitate unprecedented levels of interpersonal connections is the primary impetus for this possibility. The Metaverse has the potential to facilitate decentralized and distributed collaboration among individuals, potentially rendering centralized authority or institutions unnecessary for the execution of projects and activities.

Nevertheless, the possibility of causing disturbance also gives rise to several ethical implications. A critical issue that warrants attention is the potential for the Metaverse to amplify preexisting power differentials [2], particularly if it is monopolized by a limited number of influential actors or platforms [13]. This phenomenon has the potential to reinforce preexisting disparities in social and economic status, thereby constraining avenues for involvement and interaction.

An additional ethical consideration pertains to the potential for the Metaverse to facilitate the emergence of novel modes of governance and political structuring that may deviate from established democratic standards and tenets [3]. The Metaverse has the potential to facilitate the establishment of decentralized autonomous organizations or other types of governance frameworks that are not subject to conventional regulatory or oversight mechanisms.

In light of these apprehensions, it is imperative for policymakers and relevant parties to deliberate upon the ethical ramifications of the Metaverse and strive toward establishing a virtual realm that is characterized by fairness and impartiality [3]. The task at hand may entail the creation of novel regulatory frameworks and governance systems to uphold the principles of openness, accessibility, and democracy within the Metaverse. Additionally, it may be necessary to forestall the emergence of fresh power asymmetries or concentrations of power [11].

In the context of Metaverse development, it is imperative to consider matters of social justice and diversity [14]. To attain our goal, it may be necessary to implement measures aimed at enhancing the diversity and inclusivity of the Metaverse, while concurrently implementing strategies to mitigate the potential escalation of marginalization and exclusion within the nascent virtual realm [10].

In general, the Metaverse has the potential to cause significant disruption to established power structures and engender novel forms of social and political organization. However, it is imperative to adopt a critical and ethical perspective when considering this prospect [3]. Through this approach, it is possible to strive toward the establishment of a virtual realm that embodies the fundamental ideals and ethical standards of our wider community.

5.3 THE METAVERSE AND ITS ETHICAL IMPLICATIONS

The continued development of the Metaverse has sparked a wide range of moral concerns due to the numerous ethical dilemmas it has raised. These implications can be found in many other domains, some of which include health and wellness [8], groupthink and cognitive echochambers [7], social inequality [15], and the necessity of connecting the Metaverse with human values and beliefs [16]. These are just a few instances of the many different domains.

5.3.1 Effects of the Metaverse on Relationships, Which Include Digital Isolation and the Loss of Face-to-face Communication Skills

The Metaverse is a computer-generated simulation of a physical world that facilitates human interaction within a digital realm. The increasing prevalence of digital spaces in individuals' lives has elicited ethical considerations pertaining to social interactions and relationships. A significant issue of concern pertains to the possibility of digital isolation and its consequential effects on interpersonal connections.

A significant issue that warrants attention is the possibility of digital isolation and its effects on interpersonal connections [17]. The potential increase in Metaverse usage may lead to a decrease in real-world social interaction, potentially resulting in diminished social aptitude and difficulty forming significant relationships beyond the Metaverse. Moreover, the prioritization of virtual interactions may potentially induce a detachment from tangible reality, culminating in a predilection for digital interactions and a sense of detachment from one's physical environment.

An additional matter of concern pertains to the possibility of the Metaverse serving to fortify preexisting biases and engender echochambers, thereby curtailing the scope of exposure to heterogeneous perspectives [7]. The potential of the Metaverse to generate and perpetuate preexisting biases may exacerbate societal divisions and restrict access to varied viewpoints, resulting in a disjointed and secluded virtual environment.

In addition, the potential of the Metaverse to influence human behavior and agency via computational algorithms may have noteworthy ramifications for social inequalities [15]. With the increasing integration of the Metaverse into daily life, individuals who lack access or financial resources may face potential disadvantages in areas such as education, employment prospects, and social interaction. The potential outcome of this situation is the amplification of preexisting social disparities, particularly with regards to the availability of digital resources and technological infrastructure.

In light of the ethical considerations involved, it is imperative to reformulate the Metaverse in a manner that is more congruent with fundamental human values and principles [16]. It is imperative that the developers of the Metaverse assume accountability for ensuring that the technology is intentionally designed to cultivate positive social interactions and relationships [18]. This may involve integrating characteristics that foster inclusivity and incentivize practical involvement. Furthermore, it is imperative to involve stakeholders in the developmental stages of the Metaverse to guarantee its design is equitable and inclusive.

The ethical implications of the Metaverse's influence on social interactions and relationships are noteworthy. Developers and policymakers must exercise caution in light of the potential for digital isolation, reinforcement of preexisting biases, and exacerbation of social inequalities. In order to ensure alignment with human values and principles, as well as the promotion of healthy social interactions and relationships, it is imperative to restructure the Metaverse.

5.3.2 Consent, Ownership, and Protection of Metaverse Data

The centrality of data utilization in the Metaverse is crucial for its operation, as it facilitates customized and engaging user experiences. The ethical implications pertaining to the utilization of data in the Metaverse are intricate and diverse. The matter of consent [6] is a crucial factor to take into account. The level of comprehension among users regarding the collection and utilization of their data within the Metaverse may be inadequate, particularly in cases where such data are gathered through passive means such as sensors or other similar devices. Ensuring users possess a lucid comprehension of the utilization of their data and the capacity to regulate such usage through informed consent mechanisms is of paramount significance [19].

Data ownership is a crucial aspect to take into account [4]. Within the Metaverse, individuals possess the ability to generate and provide contributions to the platform, including virtual objects or experiences, that may hold considerable worth. It is crucial to guarantee that users retain proprietorship of their contributions and receive appropriate remuneration for their utilization within the Metaverse.

The safeguarding of data is a crucial factor to take into account [20]. It is probable that the Metaverse will amass a considerable quantity of personal information, encompassing biometric data, that may be utilized for the purpose of individual identification and tracking. The implementation of adequate measures to safeguard data from unauthorized access or misuse is crucial [7].

The utilization of data in the Metaverse raises ethical concerns that underscore the necessity of a thorough regulatory structure that tackles matters pertaining to consent, data ownership, and data safeguarding. The development of a framework is imperative to guarantee user autonomy over their data and safeguard their entitlements in the Metaverse, as stated in Ref. [5]. Furthermore, the development of the aforementioned should be conducted via a collaborative approach that encompasses the perspectives of a broad spectrum of stakeholders, such as end-users, software developers, regulatory bodies, and civil society organizations [21]. Through the resolution of these concerns, it is possible to guarantee that the Metaverse is constructed in a manner that upholds human rights and advances the welfare of all its users.

5.3.3 The Metaverse's Capacity to Transform Power Structures and Its Effects on Democracy and Social Justice, Including Digital Citizenship and Equal Access to Digital Infrastructure

The emergence of the Metaverse has the capacity to fundamentally alter established power dynamics through the establishment of novel economic and social frameworks, with digital technologies occupying a central role in this paradigm shift.

The aforementioned alteration gives rise to noteworthy ethical considerations, specifically pertaining to the principles of democracy and social equity.

A significant issue pertains to the possibility of the Metaverse aggravating preexisting disparities, encompassing those within and among societies [2–4]. Although digital technologies possess the capability to equalize access to information and resources, it is a fact that not all individuals have comparable access to these technologies, particularly in underdeveloped nations or disadvantaged communities. The aforementioned issue gives rise to apprehensions regarding the digital divide and underscores the imperative of ensuring just and fair access to digital infrastructure and services.

The Metaverse raises ethical concerns regarding digital citizenship, as highlighted in literature [14]. The concept of digital citizenship pertains to the privileges and obligations that individuals possess while utilizing digital technologies. These include, but are not limited to, the entitlement to privacy [6, 20], the entitlement to freedom of expression [14], and the duty to safeguard oneself against cyberbullying and online harassment [14]. As the Metaverse gains greater significance in our daily lives, it is imperative that we cultivate a collective comprehension of the concept of digital citizenship and the corresponding obligations that accompany this designation.

Moreover, it has been suggested that the Metaverse holds the capacity to revolutionize our understanding of democracy and governance [4]. The proliferation of digital technologies in the public sphere has led to a rising demand for novel frameworks of democratic involvement and interaction [21]. The aforementioned matter prompts significant inquiries regarding the impact of digital technologies on molding public perception and the likelihood of the Metaverse to generate novel modes of social regulation and influence [15].

In order to mitigate the ethical considerations, it is imperative to reformulate the Metaverse to more closely conform with fundamental human values and principles. The task at hand involves the creation of novel frameworks for governance and regulation that give precedence to safeguarding human rights and advancing social justice [4]. The advancement of technology necessitates the creation of novel technologies that prioritize ethical principles, including but not limited to privacy by design and transparency. This notion is supported by various scholarly sources [22].

The Metaverse's capacity to transform established power dynamics and engender novel economic and social frameworks gives rise to noteworthy ethical considerations [23–25]. Achieving alignment between the Metaverse and human values and principles requires interdisciplinary collaboration and stakeholder engagement, as emphasized by various sources. The aforementioned encompasses the necessity for fair and impartial availability of digital infrastructure, the function of digital citizenship, and the consequences for democracy and social equity.

5.4 POLICYMAKERS AND METAVERSE ETHICS

The advancement of the Metaverse necessitates the consideration of ethical implications by policy makers, as per existing literature [5]. The implementation of policies and regulations is imperative to safeguard individuals against the potential negative consequences of the Metaverse, while simultaneously ensuring that the advantages of this technology are fully realized. Within the realm of data privacy and security, it is possible to establish policies and regulations that promote responsible and transparent collection and utilization of data [6]. The aforementioned may comprise prerequisites for obtaining user consent, reducing data to a minimum, and eradicating data upon its obsolescence [1]. In addition, it is possible to formulate policies aimed at guaranteeing the secure storage of data and the implementation of suitable measures to forestall unauthorized access, as suggested by Ref. [6].

The regulation of algorithms that form the foundation of the Metaverse is an additional domain where policy interventions may prove efficacious. It has been suggested that the formulation of policies and regulations may be necessary in order to establish transparency and accountability of algorithms, as well as to prevent any form of discrimination against specific individuals or groups [5].

It is crucial to acknowledge that there exist constraints and difficulties linked to policy interventions within this framework [7]. A potential constraint of the Metaverse pertains to its global nature, whereby policies formulated by a particular country or jurisdiction may not yield desired outcomes in other regions. Moreover, the expeditious rate of technological advancement could pose a challenge for policymakers to stay abreast of the most recent advancements in the Metaverse [5].

In order to tackle the aforementioned difficulties, it is imperative to engage in interdisciplinary cooperation and involve relevant stakeholders [4]. Effective, feasible, and ethical policies will require policymakers to collaborate closely with experts in computer science, ethics, and law [5]. In addition, it is imperative for policymakers to involve stakeholders, including industry, civil society organizations, and affected individuals, in the development of policies to ensure that their needs and concerns are taken into account [7].

When it comes to addressing the ethical challenges that are brought up by the Metaverse, policymakers will play a pivotal role. It is possible for policymakers to encourage the responsible and principled use of new technologies like the Metaverse by developing policies and regulations that protect people while simultaneously encouraging the advantages of using such technologies. Effective policy interventions require the input of specialists from a variety of sectors, in addition to the flexibility to accommodate the ever-shifting nature of the technological landscape [5].

5.5 SIMULATIONS AND USER STUDY

Simulations and user studies are crucial components in the development and evaluation of smart cities in the Metaverse [26]. Simulations allow designers and researchers to create realistic digital representations of smart cities, which can be used to test and optimize different scenarios and technologies [27]. User studies, on the other hand, help to evaluate the usability and effectiveness of these smart city designs from the perspective of end-users.

Simulations can be used to test a variety of smart city features and functions, such as traffic management systems, energy-efficient buildings, and smart waste management systems [27]. These simulations can be run using various software tools such as Unity and Blender, which allow for the creation of detailed 3D models of the city and its infrastructure. The simulations can then be used to test the performance of

FIGURE 5.2 Visualization of Smart City in Metaverse.

different smart city systems under different conditions and scenarios, such as peak traffic hours, extreme weather events, or emergency situations [21, 27]. Figure 5.2 shows the visualization of smart city in Metaverse using Blender and Unity.

User studies are an important component of smart city development because they provide insights into how end-users interact with smart city technologies and infrastructure. These studies can be conducted in various ways, such as interviews, surveys, and observation studies. User studies can help to identify user needs, preferences, and pain points, which can be used to optimize smart city designs and improve user experience.

To conduct user studies, researchers can create interactive simulations that allow users to interact with different smart city systems and infrastructure [14, 27]. Users can then be asked to perform tasks, such as navigating through the city or using a particular smart city feature. Researchers can use various metrics, such as task completion time, user satisfaction, and error rates, to evaluate the effectiveness of different smart city designs [27]. Overall, simulations and user studies are essential components in the development and evaluation of smart cities in the Metaverse [26, 27]. Simulations allow designers and researchers to test and optimize different scenarios and technologies, while user studies provide valuable insights into how end-users interact with smart city infrastructure. By combining these two approaches, researchers can create smart city designs that are both effective and user-friendly.

5.6 CONCLUSION

In conclusion, this chapter has explored the ethical implications of the Metaverse as a virtual form of data-driven smart cities. The Metaverse is a hypothetical 3D network of virtual spaces being developed by Meta, which reduces the experience of everyday life to logic and calculative rules and procedures, thereby limiting human agency and autonomy. The computational understanding of human users in the Metaverse raises several ethical concerns, including privacy, surveillance capitalism, dataveillance, geosurveillance, and the potential for collective and cognitive echochambers. Furthermore, the Metaverse has the potential to exacerbate existing social inequalities and impact human health and wellness through addiction and mental health concerns. This chapter argues that there is a need to recast the Metaverse to better align with human values and principles and mitigate the pernicious effects of socially disruptive technologies.

REFERENCES

1. R. Kitchin, "The Data Revolution: Big Data, Open Data, Data Infrastructures and Their Consequences," SAGE, 2014.
2. S. Graham and S. Marvin, "Splintering Urbanism: Networked Infrastructures, Technological Mobilities and the Urban Condition," Routledge, London, 2002.
3. D. Harvey, "The right to the city," The City Reader, vol. 6, no. 1, pp. 23–40, 2008.
4. A. M. Townsend, Smart Cities: Big Data, Civic Hackers, and the Quest for a New Utopia, WW Norton & Company, London, 2013.
5. B. D. Mittelstadt and L. Floridi, "The Ethics of Big Data: Current and Foreseeable Issues in Biomedical Contexts," in The Ethics of Biomedical Big Data, Springer pp. 445–480, 2016.
6. O. Tene and J. Polonetsky, "Big data for all: Privacy and user control in the age of analytics," Northwestern Journal of Technology and Intellectual Property, vol. 11, p. xxvii, 2012.
7. J. Van Dijck, "Datafication, dataism and dataveillance: Big data between scientific paradigm and ideology," Surveillance & Society, vol. 12, no. 2, pp. 197–208, 2014.
8. C. Bellet and P. Frijters, "Big data and wellbeing," in World Happiness Report 2019, pp. 97–122, 2019.
9. N. Kshetri, "Blockchain's roles in meeting key supply chain management objectives," International Journal of Information Management, vol. 39, pp. 80–89, 2018.
10. I. Lee and K. Lee, "The internet of things (IoT): Applications, investments, and challenges for enterprises," Business Horizons, vol. 58, no. 4, pp. 431–440, 2015.
11. R. Goodspeed, "Smart cities: Moving beyond urban cybernetics to tackle wicked problems," Cambridge Journal of Regions, Economy and Society, vol. 8, no. 1, pp. 79–92, 2015.
12. R. Kitchin, "The realtime city? Big data and smart urbanism," GeoJournal, vol. 79, pp. 1–14, 2014.
13. E. Bell and T. Owen, "How Silicon Valley reengineered journalism," Columbia Journalism Review, 2017.
14. A. Vanolo, "Smartmentality: The smart city as disciplinary strategy," Urban Studies, vol. 51, no. 5, pp. 883–898, 2014.
15. M. Batty, "The New Science of Cities," MIT Press, 2013.
16. J. Bailenson, "Experience on Demand: What Virtual Reality Is, How It Works, and What It can Do," WW Norton & Company, 2018.
17. A. Wagner, "Life on the screen: Identity in the age of the internet," The Psychohistory Review, vol. 27, no. 2, p. 113, 1999.
18. J. L. Sherry, "Flow and media enjoyment," Communication Theory, vol. 14, no. 4, pp. 328–347, 2004.
19. B. J. Koops, "The trouble with European data protection law," International Data Privacy Law, vol. 4, no. 4, pp. 250–261, 2014.

20. H. Nissenbaum, "Privacy in context: Technology, policy, and the integrity of social life.", Stanford University Press, 2020.
21. G. Rudolf, C. Fertner and H. Kramar, and E. Meijers., "City-ranking of European medium-sized cities. Cent". Reg. Sci. Vienna UT, 9(1), pp.1–12.
22. S. Zuboff, "Big other: Surveillance capitalism and the prospects of an information civilization," Journal of Information Technology, vol. 30, no. 1, pp. 75–89, 2015.
23. A. Bandyopadhyay, A. Sarkar, S. Swain, D. Banik, A. E. Hassanien, S. Mallik, A. Li and H. Qin, "A game-theoretic approach for rendering immersive experiences in the Metaverse", Mathematics, vol. 11, no. 6, p. 1286, 2023.
24. A. Sihna, H. Raj, R. Das, A. Bandyopadhyay, S. Swain and S. Chakrborty, "Medical education system based on Metaverse platform: A game theoretic approach", In IEEE 4th International Conference on Intelligent Engineering and Management (ICIEM 2023), 2023, pp. 1–6.
25. P. Gupta, K. Bhadani, A. Bandyopadhyay, D. Banik and S. Swain, "Impact of Metaverse in the near 'Future'", In IEEE 4th International Conference on Intelligent Engineering and Management (ICIEM 2023), 2023, pp. 1–6.
26. A. Caragliu, C. Del Bo and P. Nijkamp, "Smart cities in Europe," Journal of Urban Technology, vol. 18, no. 2, pp. 65–82, 2011.
27. A. Kirimtat, O. Krejcar and A. Kertesz, et al., "Future trends and current state of smart city concepts: A survey," IEEE Access, vol. 8, pp. 86448–86467, 2020.

6 Game Theory in the Metaverse

Strategies, Interactions, and Implications

Renuka R. Patil
Department of Computer Science and Engineering, GITAM School of Technology

Chaithanya B N
Department of Computer Science and Engineering, GITAM School of Technology

Vamsidhar Yendapalli
Department of Computer Science and Engineering, GITAM School of Technology

Bindu Madavi K P
Department of Computer Science and Engineering, GITAM School of Technology

Geetha K
Department of Computer Science and Engineering, GITAM School of Technology

6.1 INTRODUCTION

The phrase "Metaverse" describes a virtual environment that coexists with the real world. Physical reality and persistent, augmented, or totally virtual reality (VR) have combined to form this communal, digital space. This area consists of a number of interconnected virtual worlds that coexist with our physical world. It is not just one virtual world.

Users can have real-time interactions with each other and the computer-generated environment in the Metaverse. These interactions can range from straightforward jobs like gaming and chatting to more complicated ones like exchanging virtual goods, going to virtual gatherings, and even taking part in virtual economies.

DOI: 10.1201/9781003449256-6

6.1.1 Characteristics of the Metaverse

1. **Immersive experience**: Users can interact with the environment in a life-like manner because of the immersive experience offered by the Metaverse, which is frequently made possible by VR or augmented reality (AR) technology.
2. **Persistence**: Similar to the real world, the Metaverse keeps on existing and evolving even while you're not signed in.
3. **Interoperability**: The smooth transfer of resources and data between several virtual experiences is meant by this.
4. **User-generated content**: There is more than one individual or group responsible for creating the Metaverse. Users can participate by developing digital resources, settings, and even whole universes.
5. **Economy**: With transactions, marketplaces, and a system of value for goods and services, the Metaverse has its own economy, which is frequently made possible by cryptocurrencies or other digital tokens.
6. **Accessibility**: The Metaverse should ideally be available from a variety of gadgets, such as smartphones, tablets, PCs, and specialized VR or AR technology.
7. **Social interaction**: With the ability to connect, work together, and create communities virtually, the Metaverse presents a new paradigm for social interaction.

6.1.2 Overview of Game Theory

The goal is to increase their payout, which may be thought of as the benefits or results of their interactions with other players. Game theory is based on the idea of equilibrium. When all players maintain their current strategies and none of them can unilaterally change their strategy to increase their payout, an equilibrium has been attained.The pursuit of higher payouts through player strategy is closely related to the equilibrium idea in game theory, where maintaining current strategies prevents unilateral changes that could increase individual benefits. Additionally, studying cooperation dynamics reveals how players cooperate and negotiate for mutual gain. [1].

Different categories can be used to classify games. Games that have a net reward of zero are known as zero-sum games because one player's gain is exactly balanced by another player's loss. Games with nonzero-sum outcomes permit situations in which player cooperation results in win-win outcomes. Sequential games are ones in which each player makes a move after another, possibly fully or partially knowing what was done before. In simultaneous games, decisions are made simultaneously by players who are unaware of one another's plans [2]. Repetition games entail playing the same game structure again, enabling the development of long-term strategy.

The accessibility of information can also affect how different a game is. Games with complete information are ones in which all components, including payoffs, strategies, and players, are known.

Game theory has a wide range of applications in many different disciplines, including computer science for networking and cybersecurity, political science for voting systems and diplomacy, and economics for market rivalry and price fixing [3]. Additionally, it has applications in biology, particularly in the subject of evolutionary game theory, to comprehend natural selection and survival strategies, as well as in the social sciences for the study of social dynamics and conflict resolution.

6.1.3 Justification for Metaverse Game Theory Application

Game theory may be applied effectively in the Metaverse because it is a complex, multi-agent system with a wide range of stakeholders, including people, businesses, and governments. It is the perfect setting to employ game-theoretical models for analysis and forecasting since the strategic interactions that take place within the Metaverse are not only numerous but also multidimensional.

The inherent strategic character of interactions inside this virtual world is one of the main justifications for applying game theory to it [4]. Every actor is continually making strategic decisions, whether it be individuals choosing which virtual assets to invest in, businesses forging alliances for the development of digital real estate, or governments creating legislation for virtual transactions. These choices have far-reaching effects on the ecosystem in addition to affecting their own personal payoffs. Making more informed decisions is made possible by the analytical tools that game theory offers to describe these interactions and anticipate their results.

The richness and complexity of the social, political, and even economic landscapes in the Metaverse are further strong arguments. Traditional analytical models might not be able to fully capture the complexity of such a dynamic environment. A more in-depth comprehension is provided by game theory, which emphasizes strategy, conflict, and collaboration. Nonzero-sum games, for instance, might simulate situations in which player collaboration can produce results that are mutually advantageous, like the production of public goods in online communities. This chapter of competitive interactions, such as bidding wars for virtual goods, can be aided by zero-sum games, on the other hand.

In addition, the Metaverse provides special problems that lend themselves to game-theoretical analysis, such as trust, reputation, and security [5]. For instance, the idea of a "Nash Equilibrium" can be used to explain the dynamics of trust between players, assisting in the comprehension of how harmful individuals may be discouraged inside these virtual communities. In a field that is rapidly expanding and where regulatory frameworks have not yet been fully defined, game theory can also be used to examine the relationship between innovation and regulation.

Furthermore, game theory is ideally suited for this application due to its applicability to numerous domains. Because game theory is interdisciplinary and the Metaverse connects with economics, social sciences, computer science, and even law, it offers a comprehensive approach to understanding its intricacies.

The tree diagram (Figure 6.1) that follows can be used as an introduction to grasp the fundamental concepts of game theory. The "Root" at the bottom represents the game's beginning state, where important decisions are still up in the air. This root has two branches that symbolize "Decision 1" and "Decision 2," the player's two divergent options. The player is presented with further branches after making this crucial decision, which are labeled "Strategy A, B" for "Decision 1" and "Strategy

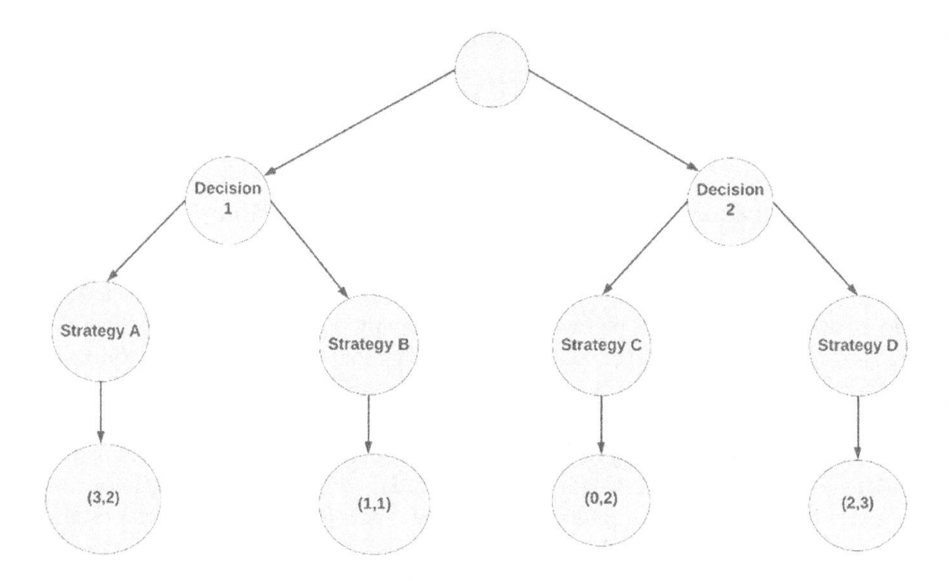

FIGURE 6.1 Tree diagram illustrating a simple game.

C, D" for "Decision 2." These tactics provide several paths to possible results or "Payoffs," which are measured in terms of rewards or advantages for the players. Selecting "Strategy A" under "Decision 1," for example, results in a payoff expressed as a pair (3,2), where "3" and "2" are the payoffs for the two players involved. As a result, the diagram graphically depicts how initial decisions lead to specific tactics, which in turn result in a variety of payoffs, demonstrating the linked complexity inherent in game theory.

6.1.4 Outline of the Chapter

This chapter tries to provide an in-depth explanation of how game theory can be applied in the Metaverse, a continuously expanding digital realm. It begins by characterizing the Metaverse as a complicated, multi-agent system, establishing the framework for the subsequent debates. The Metaverse is distinguished by its impressiveness, permanence, interoperability, and distinct economy. Following that, this chapter presents an overview of game theory, covering key concepts and types of games that serve as the theoretical foundation for the following parts. After establishing these fundamentals, this chapter articulates the reason for employing game theory to analyze the Metaverse. It investigates how the strategic, complex, and diverse character of interactions inside the Metaverse makes it an appropriate subject for game-theoretical analysis. Strategic decision-making, conflict and collaboration, and trust dynamics are identified as significant domains where game theory can provide valuable insights. This chapter finishes by summarizing these concepts and laying the groundwork for the following chapters, which will delve into specific game-theoretical models and case studies pertinent to the Metaverse. Overall, this chapter seeks to serve as a basic guide for scholars, researchers, and practitioners interested in applying game theory to comprehend the deep dynamics of the Metaverse.

6.2 THE METAVERSE AS A COMPLEX GAME

6.2.1 Introduction to the Metaverse as a Complex Game

The Metaverse is more than simply a digital cosmos; it's a complex ecology of interactions that can be modeled and understood using game theory [6, 7]. It's a place where diverse agents, ranging from individual users to corporations and governments, engage, each with their own set of goals and methods. This section goes into the intricacies of the Metaverse as a multi-agent system, the various players engaged, and the various objectives and goals that regulate behavior within this digital domain.

6.2.2 Metaverse as a Multi-agent System

The Metaverse can be thought of as a multi-agent system in which multiple sorts of agents—from human users to AI—interact within the digital environment. Each agent has some autonomy and the ability to observe its environment, make decisions based on that observation, and interact with other agent, given in Table 6.1,.

6.2.3 Players in the Metaverse (Users, Corporations, Governments)

The Metaverse is a stage with a wide cast of characters. Individual users, frequently represented by avatars, seek social connections, virtual experiences, and, in some cases, economic rewards. Corporations, on the other hand, are more concerned with maximizing profits, which they often achieve through the acquisition and trade of virtual real estate and digital assets. Governments are the new players focusing on regulation and governance.

Figure 6.2 depicts the overlap of objectives among various categories of players in the Metaverse.

Figure 6.2 clearly summarizes the intertwined goals of the Metaverse's three key players: users, corporations, and governments. Users are primarily interested with social engagement, corporations with economic gains, and governments with regulation. The overlaps between these rings signify shared goals. For example, the

TABLE 6.1
Characteristics of Agents in the Metaverse

Type of agent	Degree of autonomy	Primary objectives
Human users	High	Social interaction, economic gain
AI	Varies	Defined by programming
Corporations	Medium	Economic gain

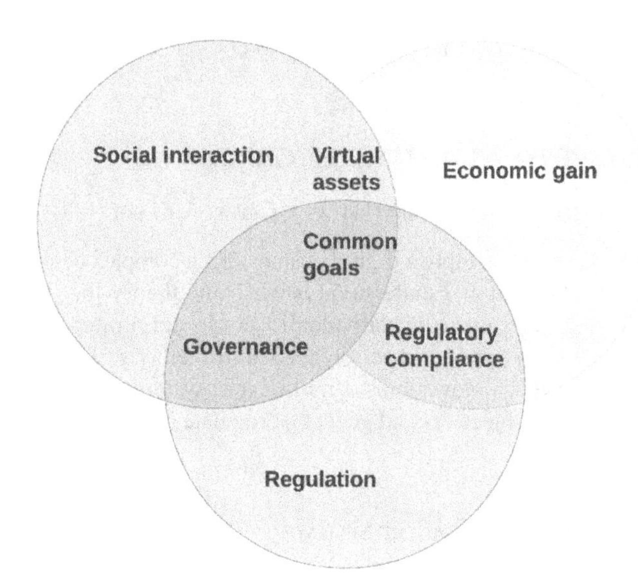

FIGURE 6.2 Overlap of objectives between different players in the Metaverse.

intersection of users and corporations is labeled "virtual assets," indicating a shared interest in digital commodities and virtual real estate. Similarly, the overlap between users and governments focuses on governance factors that affect individual behavior within the Metaverse. The corporations–governments overlap is concerned with regulatory compliance, ensuring that corporate actions in the Metaverse adhere to legal frameworks. At the center, where the three circles cross, we have "common goals," which could include universally beneficial aims like ethical practices and accessibility. The graphic serves as a concise visual guide for understanding the complicated and sometimes competing objectives of various Metaverse actors.

6.2.4 Objectives and Goals in the Metaverse (Social, Economic, Personal)

The Metaverse's goals are as varied as its participants. Some users are interested in social interaction, some in monetary gain via the trading of virtual items, and still others in personal growth or even educational gains, is given in Table 6.2. These goals do not exist in isolation; they frequently interact in intricate ways [6, 8, 9].

TABLE 6.2
Objectives and Goals in the Metaverse

Objective type	Examples
Social	Friendship, networking
Economic	Trading, investment
Personal	Education, skill development

6.2.5 Conclusion and Transition to the Next Section

As we have seen, the Metaverse is a complex game featuring a plethora of people with various goals. Understanding this complication necessitates a deep dive into strategic relationships, a subject that has a wealth of intriguing dynamics and possibilities. This acts as a transition into the following section, in which we will go into "Strategic Interactions in the Metaverse" in depth.

6.3 STRATEGIC INTERACTIONS IN THE METAVERSE

6.3.1 Introduction to Strategic Interactions

Strategic interactions in the Metaverse are a unique blend of traditional economic theory, social psychology, and game theory. The complexity of these interactions stems from the various objectives and restrictions that different parties, whether individuals, organizations, or governments, face. Understanding these relationships necessitates investigating a wide range of circumstances, from economic transactions like auctions to social dynamics like reputation management.

6.3.2 Strategic Decision-making in Virtual Economies

Strategic decision-making in the Metaverse is a multifaceted endeavor that goes beyond mere economic calculations. Players must consider not only market dynamics but also a variety of other elements such as societal conventions, regulatory rules, and even ethical considerations. A player considering investing in virtual real estate, for example, must examine not only the possibility for financial benefit, but also the social and ethical ramifications of their investment, such as how it may contribute to virtual gentrification or inequality, is given in Table 6.3.

6.3.3 Game Theory Models for Metaverse Scenarios

Game theory provides a solid foundation for analyzing strategic interactions in the Metaverse [10]. Depending on the situation, several game-theoretic models may be more or less suitable.

- **Auctions and bidding strategies**: In the Metaverse, auctions are a popular way to distribute scarce resources like exclusive digital goods or virtual property.

TABLE 6.3
Factors Influencing Strategic Decision-making

Factor	Description	Examples
Market dynamics	Supply and demand of virtual goods	Virtual real estate prices
Players behavior	Actions and reactions of other players	Trends in digital asset trading
Regulatory policies	Rules and laws governing virtual spaces	Data privacy regulations
Ethical considerations	Moral and ethical aspects	Environmental impact of blockchain transactions

- **Virtual real estate and property markets**: Because users have varying opinions and information, trading and ownership of virtual lands and properties can be complicated. Models based on game theory can be used to study how negotiations work and how prices are set in these markets.
- **Digital asset trading**: The popularity of digital assets like cryptocurrencies and nonfungible tokens has grown significantly. The strategic actions of traders and the consequent market dynamics can be better understood using game-theoretic models.
- **Social interactions and reputation management**: In the Metaverse, social dynamics are just as important as economic ones. The choice between cooperation and competition, as well as the creation and maintenance of social capital, may all be modeled using game theory.

6.3.4 Case Studies of Game-theoretic Approaches in Virtual Economies

Empirical proof of game theory's application in the Metaverse is provided via case studies from the real world. Analyses of bidding patterns in virtual auctions, tactical maneuvers in markets for digital assets, and the dynamics of cooperation and competition in virtual communities are just a few examples of the case studies that can be included, is given in Table 6.4.

6.3.5 Conclusion and Transition to the Next Section

In conclusion, this chapter of strategic interactions in the Metaverse is a rich and intricate field that combines aspects of game theory, social psychology, and economics. Anyone who wants to successfully navigate the Metaverse must have a solid understanding of this complex environment, which is not only interesting intellectually but also crucial practically. In the following part, we will delve further into the dynamics of "Conflict and Cooperation in the Metaverse," looking at how these tactical interactions may both split and bring together the many people in this virtual world.

TABLE 6.4
Notable Case Studies and Their Key Insights

Case study	Key insights	Relevance of game theory
Virtual auctions in MMORPGs	Bidding strategies, price discovery	Signaling, auction theory
Decentralized finance (DeFi)	Risk-taking behavior, smart contracts	Mechanism design
Social networks in virtual worlds	Trust, social capital	Repeated games, network theory

6.4 CONFLICT AND COOPERATION IN THE METAVERSE

6.4.1 Introduction to Conflict and Cooperation

The yin and yang of social interactions in the Metaverse are conflict and collaboration. The contrast between these two modes of interaction provides an enthralling stage for strategic exchanges. This section explores the different facets of these phenomena, from the nature of the games engaged to the Metaverse's function in collective action and diplomacy.

6.4.2 Zero-sum versus Nonzero-sum Games

In zero-sum games, one player's gains are exactly offset by another's losses, which frequently sparks conflict. Nonzero-sum games, in contrast, allow for situations in which each participant can benefit, encouraging collaboration. It's essential to comprehend this difference in order to appreciate the Metaverse's dynamics fully, is given in Table 6.5.

6.4.3 Metaverse as a Platform for Diplomacy

The Metaverse is increasingly serving as a forum for amicable conflict resolution, as seen by the prevalence of virtual embassies and diplomatic missions.

This pseudo-code provides a structured approach to understanding how diplomacy in the Metaverse could be conducted, breaking it down into manageable functions and subtasks.

Algorithm: Processes in Metaverse Diplomacy

1. Initialize:
    - Stakeholders = {Individuals, Corporations, Governments}
    - NegotiationMethods = {Virtual Embassies, Diplomatic Missions}
    - DisputeResolutionMethods = {Peace Treaties}
    - CollaborationTypes = {Joint Ventures, Alliances}
2. Function DiplomacyInMetaverse():
    2.1 Call StakeholderEngagement(Stakeholders)
    2.2 Call ConductNegotiations(NegotiationMethods)
    2.3 Call ResolveDisputes(DisputeResolutionMethods)
    2.4 Call FosterCollaboration(CollaborationTypes)

TABLE 6.5
Types of Games and Their Outcomes

Type of game	Outcome	Example
Zero-sum	Win–lose	Virtual duels
Non-zero-sum	Win–win or lose-lose	Collaborative quests

3. Function StakeholderEngagement(Stakeholders):
   3.1 For each stakeholder in Stakeholders:
      – Engage with stakeholder
      – Identify stakeholder objectives and concerns
4. Function ConductNegotiations(NegotiationMethods):
   4.1 For each method in NegotiationMethods:
      – Conduct negotiation using method
      – Record outcomes
5. Function ResolveDisputes(DisputeResolutionMethods):
   5.1 For each method in DisputeResolutionMethods:
      – Apply method to resolve disputes
      – Record outcomes
6. Function FosterCollaboration(CollaborationTypes):
   6.1 For each type in CollaborationTypes:
      – Initiate collaboration of given type
      – Record outcomes
7. End

6.4.4 Collective Action and Public Goods Provision

Collective action is another intriguing feature, particularly when it comes to delivering public goods like open-source software or neighborhood policing. Such actions frequently call for collaboration from a vast number of participants, each of whom makes a tiny contribution toward a common objective, is given in Table 6.6.

6.4.5 Virtual Communities and Group Dynamics

There are many different types of cooperation and conflict in the Metaverse's virtual communities, which range from guilds to professional networks. Game theory may be applied to these groups' dynamics to understand how trust is developed, norms are established, and conflicts are settled.

The dynamics of a virtual community are depicted in the following social network diagram (Figure 6.3). The edges between each node denote relationships or interactions, and each node represents a member of the community. This illustration can be helpful for deciphering how standards are established, how trust is developed, and how disagreements are settled within the community.

TABLE 6.6
Examples of Public Goods in the Metaverse

Public good	Description	Contributor types
Open-source software	Software freely available for all to use	Developers, users
Community policing	Shared responsibility for community security	Individual users, admins

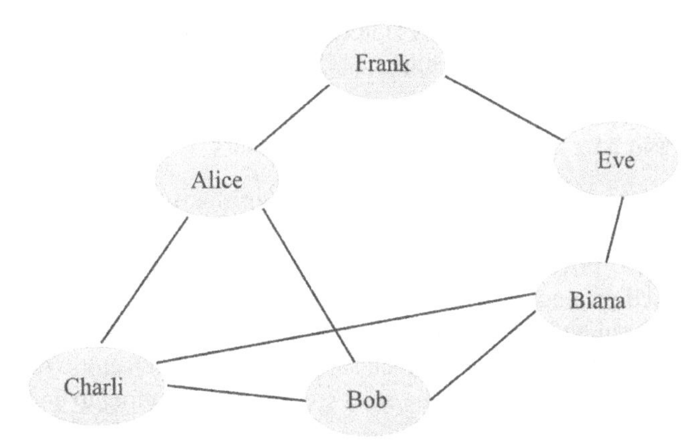

FIGURE 6.3 A social network diagram of a virtual community.

6.4.6 Illustrative Examples of Conflicts and Collaborations

Understanding the theories discussed can be improved with the use of specific instances. These instances, which range from disagreements over virtual property rights to teamwork in massively multiplayer quests, act as microcosms of more general societal problems, is given in Table 6.7.

6.4.7 Conclusion and Transition to the Next Section

The Metaverse is a rich tapestry of conflict and cooperation, each thread woven from individual choices, collective actions, and systemic rules. As we transition to the next section, we'll focus on "Virtual Asset Valuation and Auction Design," an area where these dynamics directly influence market behavior and economic outcomes.

6.5 VIRTUAL ASSET VALUATION AND AUCTION DESIGN

In the expanding domain of the Metaverse, the valuation of virtual assets and the design of auctions hold pivotal roles. Virtual assets are not merely digitized commodities; they often possess unique properties and functionalities that make their valuation complex. Various game theory models are instrumental in establishing a fair and dynamic pricing mechanism. These models help in simulating different

TABLE 6.7
Case Studies Illustrating Conflicts and Collaborations

Scenario	Nature	Outcome
Virtual property dispute	Conflict	Legal/Community ruling
Multiplayer quest	Cooperation	Shared rewards

scenarios where multiple stakeholders interact in a marketplace, thereby giving valuable insights into the assets' real value.

6.5.1 Pricing Mechanisms in Virtual Economies

In the realm of virtual economies, pricing mechanisms serve as foundational elements that regulate the exchange of virtual goods and services. Various pricing models like fixed pricing, dynamic pricing, and auction-based pricing are often employed. Fixed pricing is straightforward but may not capture the real value of an asset, which can be highly variable in a virtual setting. Dynamic pricing, on the other hand, adjusts in real-time based on supply and demand factors. Auction-based pricing lets the market determine the asset's value, often leading to competitive but fair pricing, is given in Table 6.8.

6.5.2 Game Theory Models for Valuing Digital Assets

Game theory offers valuable insights into how digital assets can be valued in a virtual economy. Models such as Nash Equilibrium can be employed to understand how rational actors would price their assets in a competitive market. These models are particularly useful in situations where multiple players have conflicting interests, and their decisions have mutual implications.

A flowchart depicting how different game theory models are applied is shown in Figure 6.4.

Figure 6.4 provides a visual representation of how various game theory models are applied to the valuation of virtual assets in the Metaverse. The main topic, "Virtual Asset Valuation," branches out into different game theory models like "Nash Equilibrium," "Zero-Sum Games," and "Bayesian Games." These models further interact with various components such as "Stakeholders," "Market Conditions," and "Asset Properties," which are essential factors in determining the value of virtual assets.

6.5.3 Designing Fair Auctions and Market Rules

The integrity of a virtual economy often depends on the fairness of its auctions and market rules. Strategies like Vickrey auctions, where the highest bidder pays the second-highest bid, can promote fairness. The rules must be transparent and enforceable to maintain trust among the participants, is given in Table 6.9.

TABLE 6.8
Comparison of Pricing Mechanisms in Virtual Economies

Pricing mechanism	Advantages	Disadvantages
Fixed pricing	Simplicity	May not capture real value
Dynamic pricing	Market-aligned	Complexity
Auction-based	Competitive	Requires active participation

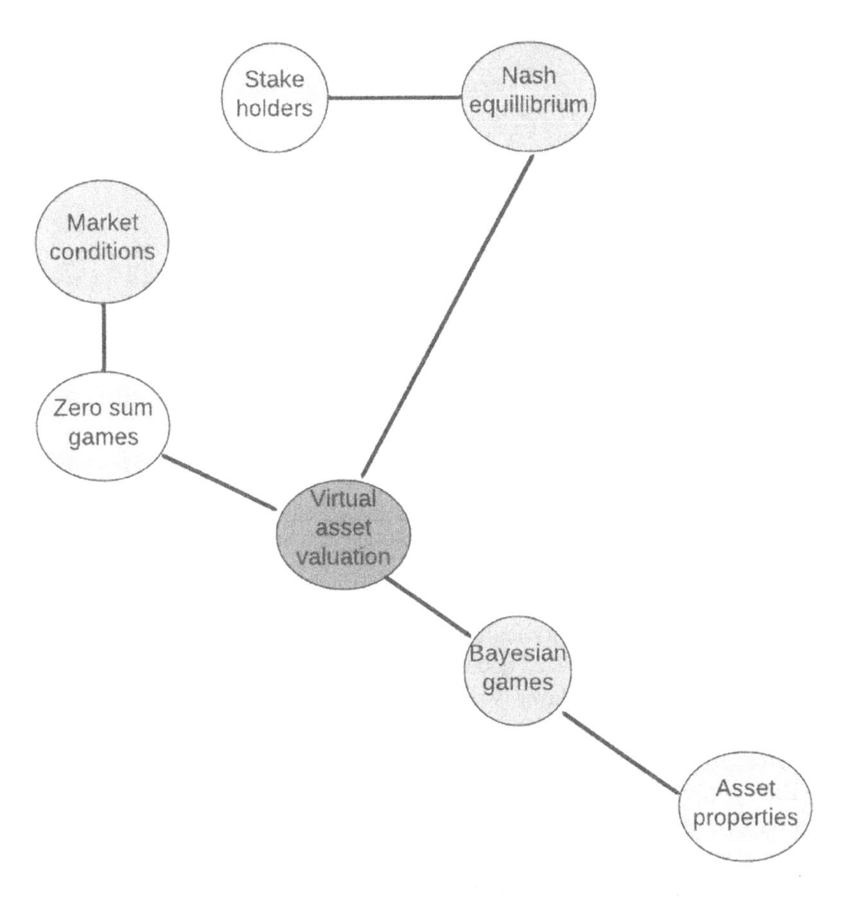

FIGURE 6.4 Game theory models applied to virtual asset valuation.

6.5.4 Ensuring Trust and Security in Virtual Transactions

Trust and security are paramount in virtual transactions. Blockchain technology, secure payment gateways, and robust authentication mechanisms are some of the techniques used to ensure a secure and trustworthy environment [11].

A layered diagram showing the different security layers is shown in Figure 6.5.

TABLE 6.9
Types of Auctions and Their Fairness Metrics

Type of auction	Fairness metric
Vickrey auction	High
Dutch auction	Medium
English auction	Moderate

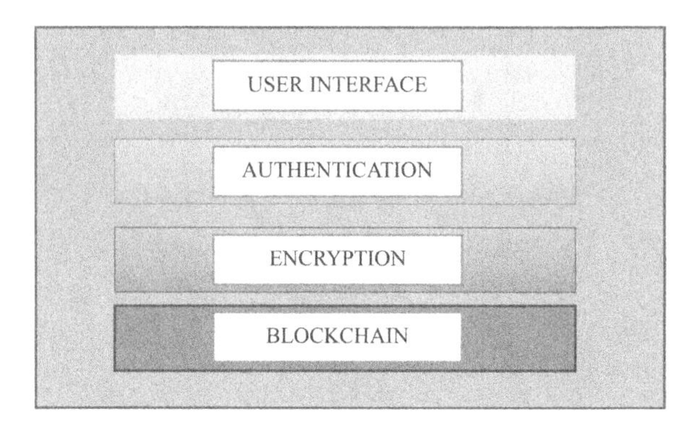

FIGURE 6.5 Trust and security mechanisms in virtual transactions.

The layered diagram above illustrates the different levels of security mechanisms involved in virtual transactions within the Metaverse. Starting from the top:

1. **User interface**: This is the front-end layer where users interact. It needs to be designed with security features like captcha and secure login to prevent phishing and unauthorized access.
2. **Authentication**: This layer is responsible for verifying the identity of users and systems. Methods like multifactor authentication are commonly used here.
3. **Encryption**: At this level, data are encrypted to safeguard it from interception and unauthorized access. SSL/TLS protocols might be employed for these purposes.
4. **Blockchain**: This is the foundational layer that ensures the immutability and transparency of transactions. Smart contracts could also be implemented at this layer for automated, secure transactions.

6.5.5 Toward Governance and Regulation in the Metaverse

While asset valuation and auction design provide the financial backbone for the Metaverse, the emerging virtual world also demands a regulatory framework [12–14]. The next section will delve into the complexities of governance and regulation in the Metaverse, exploring how game theory can be applied to create balanced regulatory strategies that protect consumers without stifling innovation.

6.6 GOVERNANCE AND REGULATION IN THE METAVERSE

6.6.1 Self-Regulation versus Government Intervention

In the Metaverse, governance can take multiple forms, ranging from self-regulation by the community to formal government intervention. Self-regulation often involves community-driven standards and decentralized mechanisms to enforce them.

TABLE 6.10
Self-regulation versus Government Intervention

Governance type	Pros	Cons
Self-regulation	Flexibility	Limited enforcement
Government intervention	Strong enforcement	May stifle innovation

While it offers more flexibility, it may lack the rigor to protect consumer interests fully. Government intervention brings statutory power into the equation, ensuring compliance but potentially stifling innovation, is given in Table 6.10.

6.6.2 Game Theory Applied to Regulatory Strategies

Game theory provides a valuable framework for understanding regulatory strategies [15–17]. By modeling interactions between regulators and entities being regulated as strategic games, one can predict the effectiveness of various regulatory approaches. Whether it's imposing fines for noncompliance or offering incentives for ethical behavior, game theory can guide the design of these mechanisms.

Figure 6.6 illustrates how various game theory models inform regulatory strategies in the governance of the Metaverse. The central node, "Regulatory Strategies," branches out into different game theory models like "Nash Equilibrium," "Prisoner's Dilemma," and "Sequential Games." These models interact with key components of regulatory strategies, such as "Regulators," "Entities," and "Compliance."

- **Nash Equilibrium** with "Regulators" signifies that optimal regulatory measures are those where no single regulator can unilaterally change its strategy to better its outcome.
- **Prisoner's Dilemma** with "Entities" implies the challenges in cooperation among the entities being regulated, particularly when noncooperation appears to provide a better outcome for individual entities but is detrimental for the collective.
- **Sequential Games** tied to "Compliance" indicates that decisions made by entities in complying or not complying with regulations can have far-reaching, sequential impacts on future interactions and regulations.

6.6.3 Balancing Innovation and Consumer Protection

Striking the right balance between fostering innovation and ensuring consumer protection is a challenging task. Overregulation could deter creativity and economic growth, while underregulation could put consumers at risk. Effective governance systems should be adaptive and data-driven, capable of evolving with technological advancements, is given in Table 6.11.

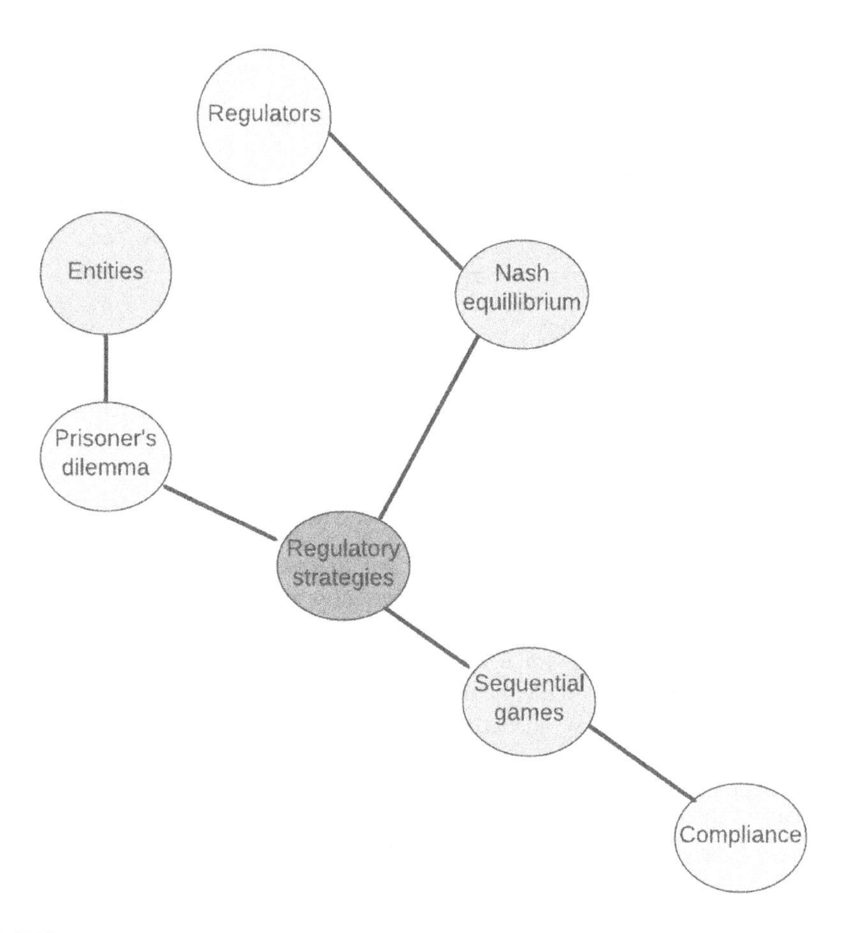

FIGURE 6.6 Game theory models in regulatory strategies.

6.6.4 Intellectual Property and Copyright Challenges

The Metaverse brings forth complex challenges related to intellectual property and copyrights [18]. From virtual real estate to digital artifacts, the lines between ownership and usage rights can blur. Regulatory frameworks need to adapt to these unique scenarios, possibly by creating new categories of intellectual property tailored for the virtual world.

TABLE 6.11
Balancing Innovation and Consumer Protection

Factor	Importance	Challenges
Innovation	Drives growth	May introduce risks
Consumer protection	Ensures safety	Can limit innovation

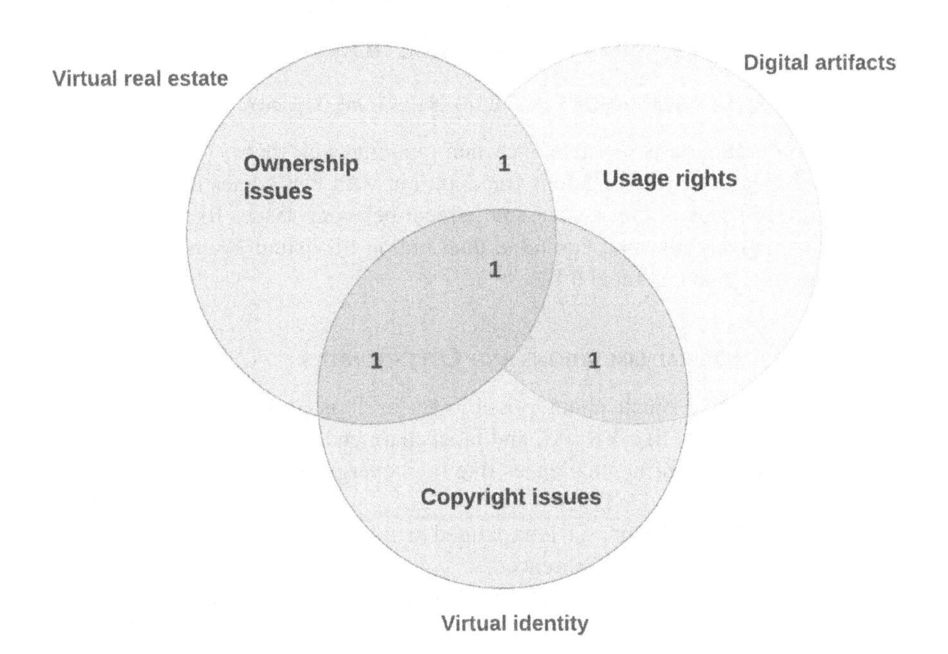

FIGURE 6.7 Intellectual property challenges in the Metaverse.

The Venn diagram offers another perspective on the intellectual property challenges in the Metaverse. Figure 6.7 consists of three main circles:

- **Virtual real estate**: The difficulties with ownership are represented by this circle. The idea of owning virtual land or property in the Metaverse is novel and complicated, frequently running afoul of established intellectual property regulations.
- **Digital artifacts**: The circle here reflects concerns with usage rights. Virtual clothing, weaponry, and other digital artifacts have their own set of rules governing how they can be used or exchanged.
- **Virtual identity**: Finally, difficulties with copyright issues are included in this circle. Copyright rules could apply to your virtual identity, which includes your avatar, name, and other unique identifiers.

It is a complicated but essential topic for governance in the Metaverse because each of these circles represents a distinct aspect of intellectual property concerns.

6.6.5 Challenges and Future Directions

It is clear that the diverse character of the virtual world necessitates similarly complex governing systems as this section on governance and regulation in the Metaverse draws to a close. The next section will delve into the overarching challenges and future directions, exploring the ethical, technological, and strategic dimensions that will shape the Metaverse's trajectory.

6.7 CHALLENGES AND FUTURE DIRECTIONS

6.7.1 Ethical Considerations in Metaverse Game Theory

Ethical considerations are a cornerstone when applying game theory to the Metaverse. The virtual space offers tremendous freedom, but with that comes the responsibility to ensure fair play, social justice, and ethical behavior. Issues like data privacy, behavioral manipulation, and equitable distribution of virtual assets need careful consideration, is given in Table 6.12.

6.7.2 Technological Limitations and Opportunities

The Metaverse is as much about possibilities as it is about limitations. While advanced technologies like VR, AI, and blockchain enable realistic and immersive experiences, they also bring challenges like high energy consumption, data storage issues, and the digital divide [19].

A SWOT analysis in Figure 6.8 is explained to outline the technological strengths, weaknesses, opportunities, and threats.

Figure 6.8 provides an overview of the technological aspects of the Metaverse, categorized into strengths, weaknesses, opportunities, and threats.

- **Strengths**: Highlighted in green, the strengths include immersive experiences made possible by advanced graphics, a global reach that transcends geographical boundaries, and real-time interaction that adds a layer of dynamism to the virtual world.
- **Weaknesses**: Marked in red, the weaknesses pinpoint areas for improvement, such as high energy consumption that raises sustainability concerns, data storage challenges, and the digital divide which can limit accessibility.
- **Opportunities**: Shown in blue, the opportunities point toward future prospects like the integration of artificial intelligence (AI) for smarter environments, the use of blockchain for enhanced security, and the potential of VR to make experiences more realistic.
- **Threats**: Indicated in orange, the threats present potential risks like data privacy, platform monopolies that could limit competition, and cyber-attacks that pose security concerns.

TABLE 6.12
Ethical Considerations

Ethical aspects	Importance	Challenges
Data privacy	High	User consent, encryption
Behavioral ethics	Moderate	Manipulation, fair play
Social equity	High	Asset distribution, access

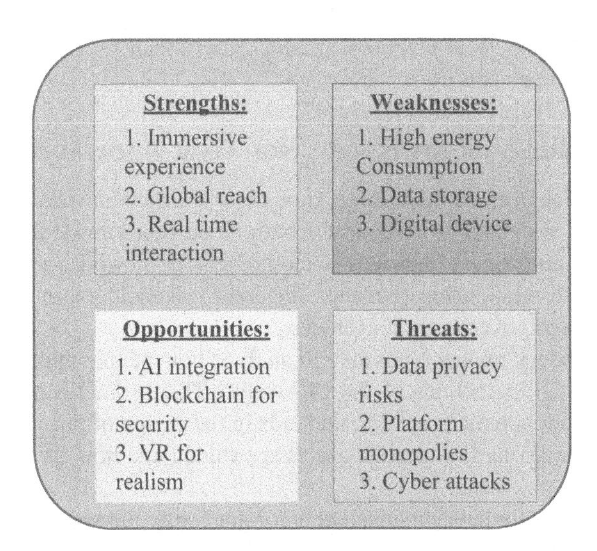

FIGURE 6.8 SWOT analysis of technology in the Metaverse.

6.7.3 Potential Risks of Strategic Behavior

Strategic behavior in the Metaverse, such as market monopolies, manipulation, and exploitation, can pose significant risks [20]. Game theory not only can help in analyzing these behaviors but also needs to account for unintended negative consequences, is given in Table 6.13.

6.7.4 Future Research Directions in Game Theory and the Metaverse

The application of game theory to the Metaverse is still a nascent field, offering a plethora of research opportunities [21, 22]. Topics like dynamic pricing models, decentralized governance, and AI-driven negotiation mechanisms are ripe for exploration.

6.8 CONCLUSION

It is obvious that the Metaverse offers a favorable environment for the application of game theory as we negotiate the complexity of obstacles and future orientations. But in order to create a thorough knowledge of this VR, an interdisciplinary approach

TABLE 6.13
Potential Risks

Risks	Consequences	Mitigation measures
Market monopolies	Reduced competition	Regulatory oversight
Manipulation	Unfair gains	Algorithmic checks
Exploitation	Inequitable distribution	Ethical guidelines

that combines insights from technology, ethics, and social sciences is also required due to its multidimensionality.

6.8.1 Summarizing the Key Insights from Game Theory in the Metaverse

The Metaverse has been studied using game theory, which has revealed a wealth of important discoveries that are crucial for both its current knowledge and its future development. Game theory provides a methodical framework for examining the intricate, strategic relationships between different stakeholders, including consumers, businesses, and governmental agencies.

First, game theory makes it possible to analyze how people make strategic decisions in the virtual economies of the Metaverse. This promotes healthy competition, prevents monopolistic practices, and aids in the design of fair markets. It offers resources for examining how virtual assets are valued and how they are exchanged or used in the virtual world.

Furthermore, game theory is invaluable for analyzing the social interactions that make up the Metaverse. It helps to comprehend reputation management, the dynamics of online communities, and the nature of disagreements and teamwork. For creating more cohesive and peaceful virtual societies, this is essential.

Third, game theory significantly influences the regulatory approaches for the Metaverse. It provides knowledge on how to balance consumer protection with innovation and how to develop just and effective governance norms. Both frameworks for state involvement and strategies for self-regulation are included.

Finally, game theory brings up crucial ethical issues including the fair distribution of resources and the players' moral behavior. It creates opportunities for investigation into the moral ramifications of activities taken in a virtual environment and strengthens the moral foundation of the Metaverse.

In essence, game theory functions as both a microscope and a construction manual, providing in-depth analysis and directing the creation of more optimized, equitable, and secure virtual worlds. Its multidisciplinary nature makes it a crucial tool for traversing the treacherous Metaverse landscapes.

REFERENCES

1. D. Fudenberg and J. Tirole, Game Theory. Cambridge, MA: MIT Press, 1991.
2. M. J. Osborne and A. Rubinstein, A Course in Game Theory. Cambridge, MA: MIT Press, 1994.
3. E. Castronova, Synthetic Worlds: The Business and Culture of Online Games. Chicago, IL: University of Chicago Press, 2005.
4. H. Zhu, "Meta-aid: A flexible framework for developing Metaverse applications via AI technology and human editing," arXiv Preprint, 2022. arXiv: 2204.01614.
5. L.-H. Lee, T. Braud, P. Zhou, L. Wang, D. Xu, Z. Lin, A. Kumar, C. Bermejo and P. Hui, "All one needs to know about Metaverse: A complete survey on technological singularity, virtual ecosystem, and research agenda," arXiv Preprint, 2021, arXiv: 2110.05352.
6. A. Bandyopadhyay, A. Sarkar, S. Swain, D. Banik, A. E. Hassanien, S. Mallik, A. Li and H. Qin, "A game-theoretic approach for rendering immersive experiences in the Metaverse," Mathematics, vol. 11, p. 1286, 2023. https://doi.org/10.3390/math11061286

7. S. Griffiths and O. Sagarra, "Game theory and social interactions in the Metaverse," Journal of Interactive Media, vol. 3, no. 2, pp. 12–27, 2020.
8. A. Sihna, H. Raj, R. Das, A. Bandyopadhyay, S. Swain and S. Chakrborty, "Medical education system based on Metaverse platform: A game theoretic approach," in IEEE 4th International Conference on Intelligent Engineering and Management (ICIEM 2023), 2023, pp. 1–6.
9. P. Gupta, K. Bhadani, A. Bandyopadhyay, D. Banik and S. Swain, "Impact of Metaverse in the near 'future'", in IEEE 4th International Conference on Intelligent Engineering and Management (ICIEM 2023), 2023, pp. 1–6.
10. R. Smith and J. Anderson, "Game theoretic models for Metaverse scenarios," Journal of Game Theory and Virtual Worlds, vol. 2, no. 1, pp. 45–60, 2021.
11. D. McDonald, "Ethical considerations in virtual asset transactions," Journal of Virtual Ethics, vol. 7, no. 2, pp. 61–76, 2021.
12. J. Sweeney, "Game theory application in virtual worlds: The case of cryptocurrencies," Journal of Virtual Economics, vol. 1, no. 1, pp. 25–40, 2019.
13. Y. Varoufakis, "Some thoughts on the political economy of virtual worlds," in Game Developers Conference, 2002.
14. T. W. Bell, "Virtual worlds, real rules," New York Law School Law Journal, vol. 49, p. 103, 2008.
15. B. Chen, C. Song, B. Lin, X. Xu, R. Tang, Y. Lin, Y. Yao, J. Timoney and T. Bi A cross-platform Metaverse data management system. In Proceedings of the 2022 IEEE International Conference on Metrology for Extended Reality, Artificial Intelligence and Neural Engineering (MetroXRAINE), Rome, Italy, 26–28 October 2022; Piscataway, NJ: IEEE, 2022, pp. 145–150.
16. S. Hollensen, P. Kotler and M. O. Opresnik, "Metaverse—The new marketing universe," Journal of Business Strategy, 2022 (ahead-of-print).
17. H. Zhu, MetaOnce: A Metaverse framework based on multi-scene relations and entity-relation-event game. arXiv Preprint, 2022. arXiv: 2203.10424.
18. J. Thompson, "Copyright and intellectual property challenges in the Metaverse," Virtual Law Review, vol. 4, no. 2, pp. 88–103, 2020.
19. T. Wu and W. Zhang, "Blockchain and game theory in the Metaverse," Journal of Blockchain Research, vol. 1, no. 1, pp. 17–32, 2022.
20. L. Zhang and Y. Zhu, "Strategic behavior and regulation in virtual economies," Virtual Economics Review, vol. 5, no. 1, pp. 33–48, 2019.
21. P. Bhattacharya, A. Verma, V. K. Prasad, S. Tanwar, B. Bhushan, B. C. Florea and A. Tolba, "Game-o-meta: Trusted federated learning scheme for P2P gaming Metaverse beyond 5G networks," Sensors, vol. 23, no. 9, p. 4201, 2023.
22. S. Van der Land, A. Schouten and F. Feldberg, "Modeling the Metaverse: A theoretical model of effective team collaboration in 3D virtual environments," Journal of Virtual Worlds Research, vol. 4, no. 3, 2011.

7 Resource Allocation in Metaverse via a Game Theoretic Auction-Based Mechanism

Anjan Bandyopadhyay
School of Computer Engineering, Kalinga Institute of Industrial Technology, Bhubaneswar, India

Priyanshu Singh
School of Computer Engineering, Kalinga Institute of Industrial Technology, Bhubaneswar, India

Ritika Pandeya
School of Computer Engineering, Kalinga Institute of Industrial Technology, Bhubaneswar, India

Roshan Singh
School of Computer Engineering, Kalinga Institute of Industrial Technology, Bhubaneswar, India

Sujata Swain
School of Computer Engineering, Kalinga Institute of Industrial Technology, Bhubaneswar, India

7.1 INTRODUCTION

Innovations in computer science have a significant impact on daily life because they transform and improve social interactions, communication, and human interaction [2]. The phrase "Metaverse" originally debuted in Neal Stephenson's science fiction novel Snow Crash, which was released in 1992 [3]. The name Metaverse is made up of the words Meta and Verse, which together imply "beyond universe" [1]. Since its introduction, a variety of concepts have come to describe the Metaverse as a computer-generated universe [4], collective space in virtuality [5], embodied internet/spatial internet [6], a mirror world and is an omniverse: a space for cooperation and simulation [8]. The concept of Metaverse is relatively new but many companies are starting to invest in this space. Facebook is one of the biggest investors in the Metaverse, and in October 2021, it rebranded itself as Meta and is completely

 DOI: 10.1201/9781003449256-7

focused on building the Metaverse. Fortnite is investing in the Metaverse through its online gaming platform. The company recently announced a $1 billion investment to create a "digital ecosystem" that will enable cross-platform play and bring players closer to the Metaverse. NVIDIA is the top provider of graphics processing units and other technology for gaming and virtual reality [26]. The business has been making investments in the Metaverse by creating technology and software that may be utilized to build more realistic and immersive virtual worlds.

An auction mechanism is a method of buying and selling goods or services through a competitive bidding process. It is a market mechanism that enables buyers and sellers to come together and determine the price of a particular item through an open and transparent bidding process as shown in Figure 7.1. Honest responses are among the most crucial characteristics of an auction mechanism [9]. The auction process seeks to optimize the social welfare of resource suppliers while also maximizing their financial gain [10]. In the Metaverse, virtual assets like digital art, virtual property, and rare virtual goods that are challenging to value in conventional markets are frequently sold through auctions. The auction can take place in a virtual space, where buyers and sellers can participate through their avatars, or it can be conducted entirely online through a digital platform. The following is a description of the remaining portions of our article. Section 7.2 illustrates the earlier studies in the relevant field. The system model and formulations are then shown in Section 7.3. Then, in Section 7.4, we suggest a payment algorithm and a

FIGURE 7.1 Auction process.

heuristic resource allocation method for an online auction mechanism. We demonstrate the experimental findings and performance analysis of the suggested technique in Section 7.5 using graphs. The conclusion and future course of our research are mentioned in Section 7.6.

7.2 LITERATURE REVIEW

With the advent of the digital age, a large portion of the world migrated online, driving a long-term digital revolution. Several companies changed to online business models, and many workers began working from home. The vast majority of services in our area now have an internet presence [24]. Metaverse focuses on the virtual experience to the next level by creating a virtual environment. The Metaverse is recognized as the next evolution of the internet and is developed in such a way as to bring new opportunities for us to create our personal space virtually where we can facilitate commerce, entertainment, and communication which leads us to new opportunities for human connection and creativity. Neal Stephenson, an American writer in the year 1992, published his science fiction book "*Snow Crash*" which introduced the world to Metaverse [3, 27, 28, 29]. In this novel, people use digital avatars of themselves to explore the online world, or we can say it is a way they can escape reality in a virtual world [11]. Like, imagine a virtual shared universe, where we together can create some unparalleled extraordinary opportunities that completely redefine the very way we used to live, play, or do our work. Numerous online events supplant in-person social gatherings and events, such as the virtual beginning held by Berkeley in 2021 and also Fortnite in 2019 hosted a virtual concert that was watched by over 10 million people [12]. Game developers have developed naturalistic virtual worlds which are very similar or way more undistorted to learn, enjoy, and even earn money. It's a giant step toward something futuristic. The Metaverse will be the new working or living space for people around the world. It will replace their usual hangout or working place. As it develops, most people will spend their time in this virtual environment learning new things and doing useful things. From an investing viewpoint, it's very important and we surely cannot ignore this. The Metaverse will transform everything we usually do including learning, playing, or doing our work. It's the next big revolution. In several disciplines, including communication networks, traffic networks, and electrical power systems, the problem of resource allocation among some organizations has been a hot topic for research. Auction mechanisms have several advantages that make them useful tools for allocating resources efficiently and fairly [13].

Since the Metaverse is still developing, little has been done in the area of allocating wireless resources to it. Several network and communication issues arise when we build a digital twin in the Metaverse [14]. The information gathered from Internet of Things (IoT) devices must be used by a virtual service provider (VSP), which is used to create a new virtual representation of the real world and to render the Metaverse [15]. With limited bandwidth and a large amount of data being captured from IoT devices (such as photos and videos), VSP becomes less effective at retrieving all the data from the real world. Hence, the Metaverse needs to render digital data quickly while limiting transmission latency. We suggest IoT devices use semantic

FIGURE 7.2 Truthful auction mechanism.

information extraction methods to address this issue and reduce the amount of transferred data in the wireless channel.

We created a trustworthy auction mechanism, as shown in Figure 7.2, in the Metaverse to assist them in obtaining the necessary data from the devices, as many IoT devices are interested in selling their information or data to the VSP.

7.3 SYSTEM MODEL AND PROBLEM FORMULATION

An auction mechanism is a system in which items or assets are sold to the highest bidder. In a Metaverse, an auction mechanism could be used to buy and sell virtual goods or devices [15]. Here, we are using the second-highest bidding theorem based on the VCG auction mechanism. In the Metaverse, we can use it to allocate virtual goods or devices we need to sell or buy. It is intended to persuade bidders to disclose their actual assessment of an object and to award the item to the bidder who places the highest value on it. The highest bidder receives the item after bidders submit their offers. However, the winning bidder does not pay the same amount as their bid; instead, the price is established by the other bidders. The amount paid by the winner is specifically equal to the total of the second-highest bids made by every other bidder [16]. This auction mechanism is designed to incentive bidders to bid truthfully because a bidder who bids their true value for the item will not be penalized if they win the auction and will not be rewarded if they lose the auction. This can result in a more efficient allocation of resources, as the item is allocated to the bidder who values it the most, rather than the bidder who is willing to pay the most.

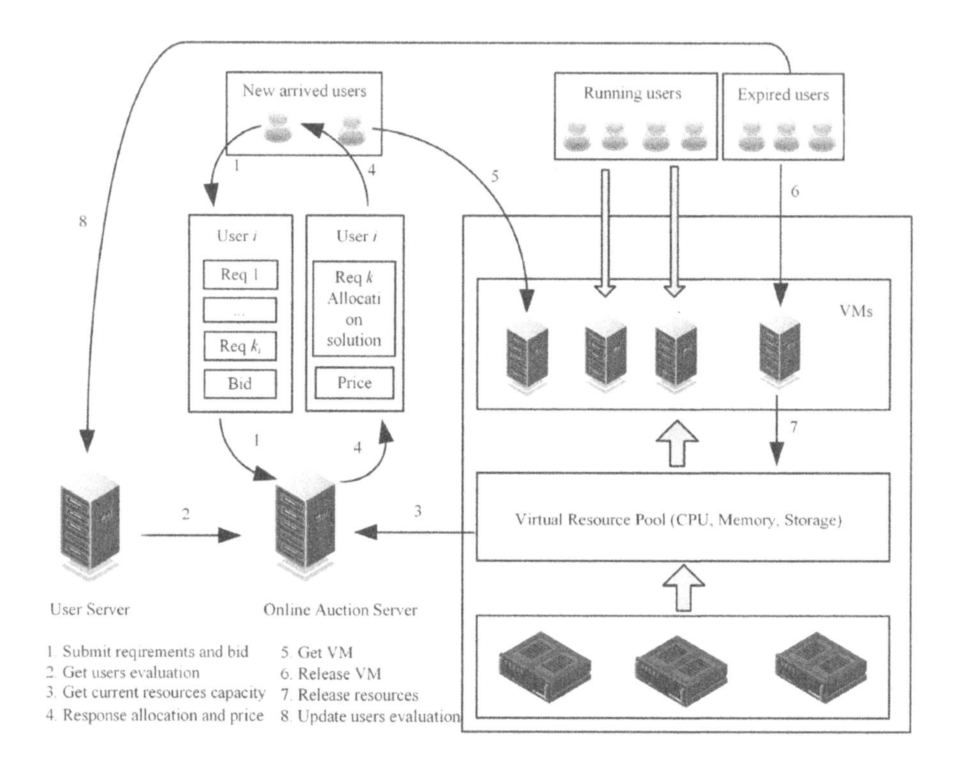

FIGURE 7.3 Framework of online auction in Metaverse.

In the Metaverse, we can use the second-highest bidding theorem to allocate devices fairly and efficiently. For example, the winning bidder would pay an amount equal to the total of the second-highest bids made by all other bidders to distribute virtual gadgets in the Metaverse using this auction method (Figure 7.3). This would ensure that the device is allocated to the bidder who values it the most, while also encouraging bidders to bid truthfully [17].

1. Allocation formula:

   The allocation formula is used to determine which bidder is allocated the item. It is based on the bids of all bidders and their valuation of the item.

   Let there is n number of the bidder, the allocation formula is:

$$A(j) = \text{argmax}\{\text{sum of values of all bidders except } j\}, \quad \text{where } j = 1, 2, \ldots, n$$

   This formula states that the item is allocated to the bidder who has the highest sum of the values they place on the item, where the sum is calculated by adding up the values of all bidders except for bidder j.

2. Payment formula:

   The payment formula is used to determine the price paid by the winning bidder. It is based on the bids of all bidders and their valuation of the item.

For this auction with n bidders, the payment formula is:

$$P(j) = \text{sum of the second} - \text{highest bids of all bidders except } j$$

According to this formula, the winning bidder must pay the total of all other bidders' second-highest bids. This encourages bidders to place honest bids and guarantees that the winning bidder would pay the amount that would have been paid if they had not entered the auction.

3. Cost formula:

To calculate the cost that each bidder will be charged in a VCG auction, the following formula can be used:

$$\text{Cost}_i = \sum_{j=1,\, j \neq i}^{n} V_j(x_{-i}, x_i) - \sum_{j=1,\, j \neq i}^{n} V_j(x_{-i})$$

where:

Cost_i is the cost to bidder i.
V_j is the valuation function of bidder j.
x_{-i} is the allocation of goods to all bidders except bidder i.
x_i is the allocation of goods to bidder i.

In other words, the cost to bidder i is equal to the total loss in value to all other bidders that result from bidder "i" participating in the auction. This ensures that bidders are charged a fair price for the items they receive and that the auction is efficient in allocating goods to the bidders who value them the most.

7.4 PROPOSED MECHANISM

In this section, we have proposed an auction mechanism for selling virtual assets on Metaverse. Here, we introduce the weighted VCG auction mechanism (Vickrey–Clarke–Groves) auction. It is a type of auction designed for the sale of multiple goods or assets, where bidders have different valuations for each item. In a VCG auction, each bidder submits a bid for each of the goods or assets being sold. The auctioneer then determines the winning allocation of goods or assets that maximizes the total value generated by the auction. Each bidder pays an amount equivalent to the harm they cause to the other bidders by winning the auction [18].

The VCG mechanism provides a way for online resource allocation and calculates the payment price of the resources. Resource allocation is the first issue that must be solved when developing an auction process because it is necessary for determining the payout price. Maximizing social welfare must be considered while calculating the payment price [19].

To determine whether the kth need of user i is allocated, we create z_{ik}, where $z_{ik} = 1$ signals that the demand is allocated at time t_i and $z_{ik} = 0$ otherwise.

$$z_{ik} = \begin{cases} 1, & \text{allocated at time } t_i \\ 0, & \text{otherwise} \end{cases} \tag{7.1}$$

The indicator parameter δ*ikt* in formula guarantees that user i will consume the resources during $[t_i, t_i + e_{ik}]$ if the kth need of user i is successfully allocated. We define S_w as the general social welfare and describe the resource allocation issue in an integer program for Metaverse computing.

Objective:

$$S_\omega = \max\left[\sum_{i\in U}\sum_{k=1}^{k_i}\left(\omega_i b_i - \sum_{r=1}^{A} s_{ikr}\upsilon_r\right)z_{ik}\right] \tag{7.2}$$

Subject to:

$$\sum_{i\in U}\sum_{k:t\in(t_i,t_i+e_{ik})} S_{ikr}Z_{ik}\delta_{ikt} \le C_r;\quad \forall r\in R;\quad \forall t = 1,2,\ldots,T \tag{7.2a}$$

$$\sum_{r=1}^{A} s_{ikr}z_{ik}\upsilon_r \le \omega_i\cdot b_i,\quad \forall i\in U,\quad \forall k\in K_i \tag{7.2b}$$

$$\sum_{k=1}^{k_i} z_{ik} \le 1,\quad \forall i\in U \tag{7.2c}$$

$$z_{ik} = \{0,1\} \tag{7.2d}$$

where S_w stands for overall social welfare, which is calculated as the weighted average of all user requirements bids w_ib_i less the associated cost $\Sigma_{r=1}^{A} S_{ikr}\cdot V_r$. According to formula (2a), the $r = 1$ resource allocation at any time t, $t \in T$, cannot exceed the capacity of any type of resource. The formula states that the cost of each demand for user i must be lower than his weighted valuation w_ib_i (2b). Only one of the needs supplied by user i can be allocated, according to formula (2c). We assume that each user has the same number of requirements, $k_i = K$, $i \in U$, without losing generality. The solution is given by an $n*k$ matrix, X. The aforementioned problem can be compared to an MKP, which is strongly NP-hard.

Theorem 1. Allocation is a feasible solution of formula (2).

Proof. According to the code (lines 7–11), every resource can be allocated before being used, which satisfies the formula's conditions (2a). Each user's necessary cost must be less than the valuation bi, which is guaranteed by the code's (lines 1–6) calculation (2b). Line 18 demonstrates that to meet the formula, each user has exactly one demand that needs to be allocated (2c). Line 15 states that the provided solution is an integer set and fulfills the given condition (2d). Formula (2a) is sufficient to satisfy the solution X_t of each time because allocation is event-driven (2d). The union of the solutions X_t, which also follows formula (2a), represents the whole solution X in time T (2d). A workable answer to the formula is Solution X (2) [20].

Algorithm 1: (RESOURCE_Allocation)

Input:

The current time: t;

The user specifications that have been provided at the current time t: β;

The users who have been allocated resources but whose jobs are not completed at the current time t: β_0;

The capacity of each sort of resource that is still available at time t: $R_t = (R_1^t, R_2^t, \ldots, R_m^t)$, The users' evaluation weighted values $W = (w_1, w_2, \ldots, w_n)$

Output:

The allocation solution of this allocation process: D The social welfare in this allocation process: Aswt

Rt ← recycle_resources (Rt, β_0, t)

for each i ϵ {i | $\beta_i \subseteq \beta$}, k ← 0 to K do

Calculate the resource density Yik

endfor

for D ← descend_sort(Yik)

for each i ϵ {i | $\beta_i \subseteq \beta$}, k ← 0 to K, based on non-increasing order D, do

for each r ← 1 to m do

if ($R_r^t - S_{ikr}$) ≤ 0 or $\beta = \phi$

return X_t, Uswt;

end if

end for

for each r ← 1 to m do

$R_r^t \leftarrow R_r^t - $ Sikr

end

for Zik ← 1

Xt ← Xt ∪ {i};

Uswt ← Uswt + bi;

$\beta \leftarrow \beta \setminus \beta_i$/* shows that user i has already received a resource allocation*/

end for

return Xt, Uswt

Theorem 2. The allocation algorithm has O(nkm) time complexity.

Proof. The complexity of the allocation algorithm is $O(nkm)$ because k requirements with m resource types and n users can be considered [20].

Theorem 3. Payment's algorithm complexity is $O(n2km \log b\text{max})$, where $b$max is the highest possible value of b_i.

Proof. The payment algorithm must decide the payment amount for the successful user because the worst-case scenario involves n users Xt, who will be chosen from

Algorithm 2. As line 14 of the function calls the allocation algorithm, which has an algorithm complexity of *O*(*nkm*), the complexity of using the duality to solve *b*max is log *b*max. Hence, the payment algorithm's complexity is O (*n*2*km* log *b*max).

Algorithm 2: (RESOURCE_Allocation_PAYMENT)

Input:

The output of Algorithm 1: *Xt*

The capacity of each type of resource left at time *t* before allocation: R_0^t

The user's requirement that have been submitted at the current time *t*: β;

The users who have been allocated resources but whose jobs are not completed at the current time *t*: β_0

The users weighted values $W = (w_1, w_2, \ldots, w_n)$

Output:

The payment solution at the current time *t*: P *t*

1. $\varepsilon \leftarrow$ 10−6
2. for each i ϵ {i | zik=1, zik ϵ Xt } do
3. pi ← bi; $p_i^0 \leftarrow 0$; bi ← $\dfrac{pi + p_i^0}{2}$
4. While (| pi - p_i^0|) do
5. U*swt, X*t ← Allocation (β, β_0, R_0^t, W), wi
6. if zik=1, zik ϵ Xt
7. pi ← bi; bi ← $\dfrac{pi + p_i^0}{2}$
8. else
9. p_i^0 ← bi; bi ← $\dfrac{pi + p_i^0}{2}$
10. end if
11. end while
12. $pi \leftarrow \dfrac{p_i}{w_i}$
13. P t ← P t ∪ pi;
14. end for
15. return P t

Truthful

Theoretically, a truthful auction process should lead to each party employing a true value bid bi as their dominating tactic. Hence, users cannot increase their benefits by false valuations. We believe that users will submit accurate needs for a virtual resource auction, but they might also submit inaccurate valuations. Once the user has allocated resources successfully, he will use those resources for a prolonged time [21].

Theorem 4. The allocation process is truthful.

Proof

According to our presumption, user i won the resource allocation and had his k′th criteria met. His last payment price converges to, according to the Payment algorithm,

$$p_i = \min\left(f_{jk}\right) \cdot \sum_{r \in R} \left(\frac{S_{ik'r}}{c_r} \cdot h_r \right) \cdot e_{ik'} \cdot \frac{1}{\omega_i}, \quad j \in A(\theta), \quad k \in K_i$$

7.5 RESULT ANALYSIS

In this chapter, we propose an online virtual resource auction method using user-evaluated value and cost as the foundation. Here, we compare the CPLEX-calculated OPTIMAL solution with the RESOURCE_Allocation method, which takes less time to run and can guarantee the integrity of the auction process. RESOURCE_Allocation performed admirably in terms of resource usage as well. Even though a user can only input one criterion, RESOURCE_Allocation still has several advantages over the old approach. RESOURCE_Allocation also takes into account the interests of resource producers and users [20]. For this online virtual auction mechanism, the RESOURCE_Allocation algorithm is programmed using the R programming language.

Here, the test samples are divided into three groups according to their range: mini scale (sample 1–2), average scale (sample 3–4), and wide scale (sample 5–6) based on the total number of users, which impersonate actual auction environments.

The initial set of six samples is listed in Table 7.1 as a prerequisite for all further investigations. Each experiment proves the RESOURCE_Allocation algorithm and the OPTIMAL algorithm theory by applying a varying number of needs k while keeping the same resource capacity (CPU: 40, Memory: 4000 GB, Storage: 40,000 GB). The social welfare, execution time revenue, resource usage, and served consumers of the two algorithms are contrasted in Figure 7.4, Figure 7.5, and Figure 7.6, respectively.

Figure 7.4 compares the RESOURCE_Allocation algorithm and the ideal algorithm's revenue. The most effective algorithm uses the weighted VCG mechanism to determine the payment price.

TABLE 7.1
6 Samples in Experiment

Sample	Total users (n)	Total requirements (k)	Resource capacity (CPU, memory, storage)
1	100	5	(50, 5000, 50,000)
2	100	20	(50, 5000, 50,000)
3	1000	5	(50, 5000, 50,000)
4	1000	20	(50, 5000, 50,000)
5	10,000	5	(50, 5000, 50,000)
6	10,000	20	(50,5000,50,000)

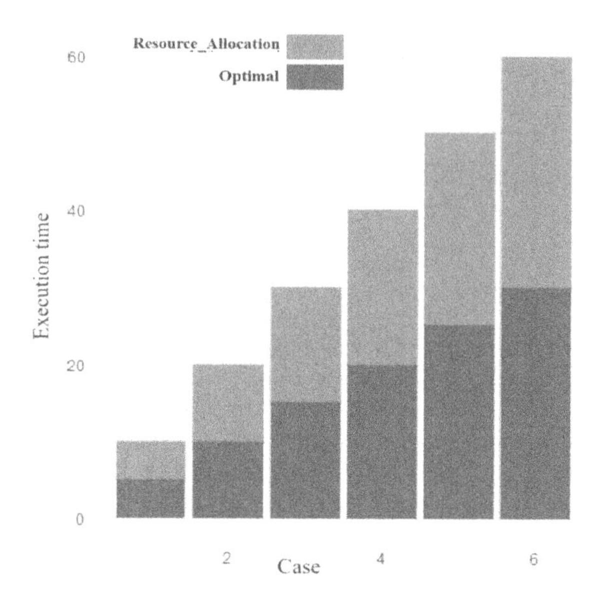

FIGURE 7.4 Bidder versus revenue.

Figure 7.5 compares the RESOURCE_Allocation algorithm with the OPTIMAL algorithm from the viewpoint of social welfare. The resource provider will benefit more from the user's submission of 20 groups than from the user's submission of only 5 groups.

Figure 7.6 compares the RESOURCE_Allocation algorithm and the OPTIMAL algorithm's processing times. RESOURCE_Allocation outperforms the ideal

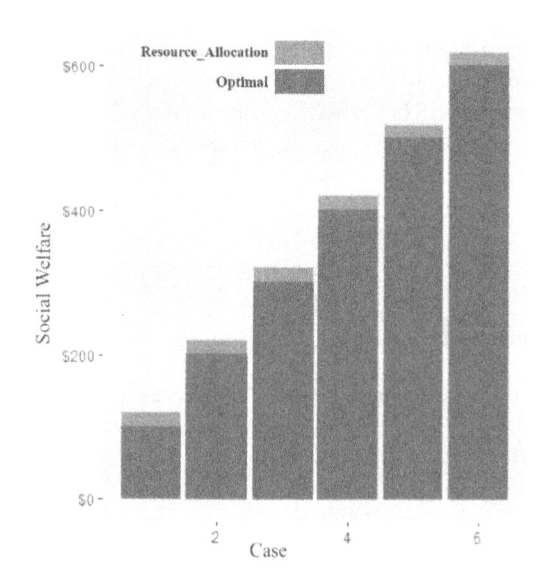

FIGURE 7.5 RESOURCE_Allocation versus OPTIMAL: Social welfare.

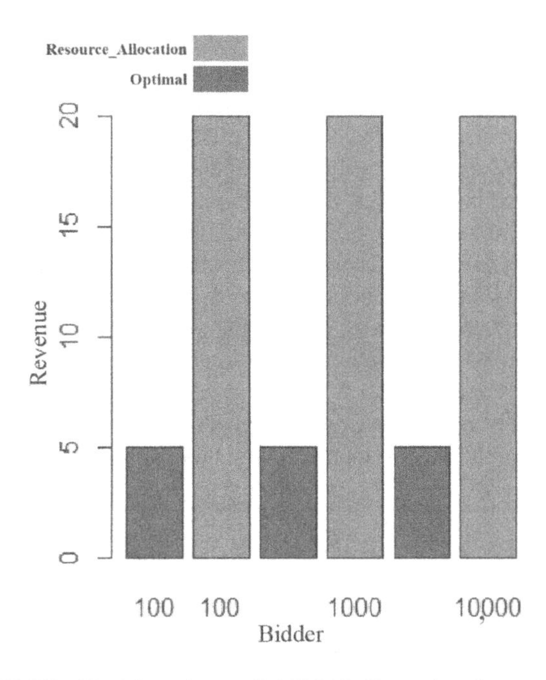

FIGURE 7.6 RESOURCE_Allocation vs OPTIMAL: Execution time.

algorithm in speed. The experimental results in Figures 7.4, 7.5, and 7.6 demonstrate that the RESOURCE_Allocation mechanism can resolve online problems including multiple requirements, multiple resource allocation, and pricing and that it has favorable execution times and financial outcomes.

7.6 CONCLUSION AND FUTURE WORK

Based on user evaluations of value and cost, we propose a Metaverse-based online virtual resource auction mechanism. Allowing users to submit a variety of environments in online auctions may enhance social welfare and increase the effectiveness of online auctions, according to theoretical studies and experimental results [22].

The RESOURCE_Allocation method takes less time to run than the CPLEX-determined OPTIMAL solution and can guarantee the integrity of the auction process. In terms of resource usage, RESOURCE_Allocation also performs admirably. Even if a user is only permitted to submit one demand, RESOURCE_Allocation still has definite advantages. The Metaverse's online auction systems provide a singular chance to design captivating and immersive experiences for both buyers and sellers. Here are some potential areas for future work in this field.

Effective virtual asset management will become more crucial as its significance grows. Future research in this field might concentrate on creating interfaces and tools to aid users in managing and tracking their virtual assets, particularly those bought through online auctions.

Verifying bidders' identities is crucial in an online auction setting to avoid fraud and guarantee fair bidding. The use of biometric authentication or blockchain-based identification solutions, for example, could be investigated in future studies as novel approaches to confirming bidder identities in the Metaverse [23].

Online auction mechanisms in the Metaverse could benefit from improved auction designs that are better suited to the unique characteristics of virtual assets. Future work could explore different auction designs.

Cryptocurrencies are being utilized more often in online auctions, and their incorporation into Metaverse auction processes may create new opportunities for trading virtual goods. In the future, researchers might look into how to incorporate cryptocurrencies into Metaverse auction platforms and create new financial products that make use of this integration.

Overall, the Metaverse's online auction processes present a rich and fascinating field for future research and development, with numerous chances for creativity and experimentation.

REFERENCES

1. Lidar-Based Real-Time Mapping for Digital Twin Development, July 21. Evan Brock, 2021 IEEE International Conference on Multimedia and Expo (ICME), Shenzhen, China, 2021, pp. 1–6, doi: 10.1109/ICME51207.2021.9428337.
2. I. Hatzilygeroudis, "Metaverse," Encyclopedia, vol. 2, no. 1, pp. 486–497, 2022. https://doi.org/10.3390/encyclopedia2010031
3. J. Joshua, "Information bodies: Computational anxiety in Neal Stephenson's snow crash," Interdisciplinary Literary Studies, vol. 19, no. 1, pp. 17–47, 2017.
4. A. Bruun and M. L. Stentoft, "Lifelogging in the wild: Participant experiences of using lifelogging as a research tool," in INTERACT, 2019.
5. W. Burns III. Everything You Know about the Metaverse Is Wrong? August 26, 2017. https://www.linkedin.com/pulse/everything-you-know-metaverse-wrong-william-burns-iii
6. K. Chayka. Facebook Wants Us to Live in the Metaverse, 2021. https://www.newyorker.com/culture/infinite-scroll/facebook-wants-us-to-live-in-the-metaverse
7. Nvidia omniverseTM platform, Aug 2021. https://nvidianews.nvidia.com/news/nvidia-brings-millions-more-into-the-metaverse-with-expanded-omniverse-platform
8. L. Mashayekhy, M. M. Nejad and D. Grosu, "APTAS mechanism for provisioning and allocation of heterogeneous cloud resources," IEEE Transactions on Parallel and Distributed Systems, vol. 26, no. 9, pp. 2386–2399, 2015.
9. J. Zhang, N. Xie, X. Zhang and W. Li, "An online auction mechanism for cloud computing resource allocation and pricing based on user evaluation and cost," Future Generation Computer Systems, vol. 89, pp. 286–299, 2018. https://doi.org/10.1016/j.future.2018.06.034
10. T. Huddleston (November 3, 2021). This 29-Year-Old Book Predicted the "Metaverse"—And Some of Facebook's Plans Are Eerily Similar. CNBC. https://www.cnbc.com/2021/11/03/how-the-1992-sci-fi-novel-snow-crash-predicted-facebooks-metaverse.html
11. The Metaverse's moment in time | Cvent Blog (n.d.). https://www.cvent.com/uk/blog/events/metaverses-moment-time
12. Wikipedia Contributors, Auction Theory. Wikipedia, January 29, 2023. https://en.wikipedia.org/wiki/Auction_theory
13. L. U. Khan, M. Guizani, D. Niyato and M. Debbah, Metaverse for Wireless Systems: Architecture, Advances, Standardization, and Open Challenges, January 26, 2023.

https://www.researchgate.net/publication/367529606_Metaverse_for_Wireless_Systems_Architecture_Advances_Standardization_and_Open_Challenges
14. Semantic Information Market for the Metaverse: An Auction Based Approach, April 2022. Cornell University (cs. GT). https://doi.org/10.48550/arXiv.2204.04878
15. Varian, Hal R., and Christopher Harris. 2014. "The VCG Auction in Theory and Practice." American Economic Review, 104 (5): 442–45. DOI: 10.1257/aer.104.5.442
16. J. Zhang, M. Zong and W. Li, "A truthful mechanism for multibase station resource allocation in Metaverse digital twin framework," Processes, vol. 10, p. 2601, 2022.
17. H. Varian and C. Harris, "The VCG auction in theory and practice" The American Economic Review, vol. 104, 2014. 10.1257/aer.104.5.442.
18. An Online Auction Mechanism for Cloud Computing Resource Allocation and Pricing Based on User Evaluation and Cost. Elsevier, July 2018. https://doi.org/10.1016/j.future.2018.06.034
19. "Truthful online auction toward maximized instance utilization in the Cloud", IEEE/ACM Transactions on Networking, vol. 26, no. 5, October 2018.
20. S. Kim, Auction, Learning and Bargaining based Control Scheme for Edge-assisted Metaverse System, Nov 2022. https://doi.org/10.1016/j.comnet.2022.109462
21. H. Reda (December 21, 2022). Digital Identity in the Metaverse—CryptoStars. Medium. https://blog.cryptostars.is/digital-identity-in-the-metaverse-7dd948c4d4b0
22. B. Marr, How the COVID-19 Pandemic Is Fast-Tracking Digital …, March 17, 2020. https://www.forbes.com. https://www.forbes.com/sites/bernardmarr/2020/03/17/how-the-covid-19-pandemic-is-fast-tracking-digital-transformation-in-companies
23. A. S. Gillis, Graphics Processing Unit (GPU), November 8, 2023. https://www.techtarget.com/searchvirtualdesktop/definition/GPU-graphics-processing-unit
24. A. Bandyopadhyay, A. Sarkar, S. Swain, D. Banik, A. E. Hassanien, S. Mallik, A. Li and H. Qin, "A game-theoretic approach for rendering immersive experiences in the Metaverse", Mathematics, vol. 11, no. 6, p. 1286, 2023.
25. A. Sihna, H. Raj, R. Das, A. Bandyopadhyay, S. Swain and S. Chakrborty, "Medical education system based on Metaverse platform: A game theoretic approach", in IEEE 4th International Conference on Intelligent Engineering and Management (ICIEM 2023), 2023, pp. 1–6.
26. P. Gupta, K. Bhadani, A. Bandyopadhyay, D. Banik and S. Swain, "Impact of Metaverse in the near 'future'", in IEEE 4th International Conference on Intelligent Engineering and Management (ICIEM 2023), 2023, pp. 1–6.

8 Metaverse Using Game Theory in Healthcare

Simandhar Kumar Baid
School of Computer Science and Engineering, Kalinga Institute of Industrial Technology, Bhubaneshwar, India

Anjan Bandyopadhyay
School of Computer Science and Engineering, Kalinga Institute of Industrial Technology, Bhubaneshwar, India

Parivesh Srivastava
School of Computer Science and Engineering, Kalinga Institute of Industrial Technology, Bhubaneshwar, India

Sujata Swain
School of Computer Science and Engineering, Kalinga Institute of Industrial Technology, Bhubaneshwar, India

8.1 INTRODUCTION

In today's era, the availability of digital technology has transformed various industries, including the healthcare sector. Fusion and game theory play an important role in the Metaverse. The Metaverse can be defined as a virtual shared space where two realities are merged, physical and digital. Through the Metaverse, it will be easier for people to interact with each other and other digital objects in real-time through digital representations [1]. Game theory refers to a mathematical concept that includes a number of techniques and then uses it to find situation types of multiple users. It can also be defined as a system that helps determine which actions can achieve the best or best solution for a situation between independent or competing players.

The application of Metaverse is rapidly increasing in the field of healthcare. It is creating an impact by offering the best and optimal solutions that can improve patient care, medical training, and research work. Communication with patients with the help of visiting them virtually through Metaverse reduces the need for physical appointments. Metaverse enables the interaction between patients and doctors in such a way that it becomes easier for the patient to discuss symptoms, diagnoses, and even receive prescriptions [2]. This method can be most effective in the field of remote or underserved areas [3]. Medical professionals and students can use the platform of Metaverse to practice surgical procedures, diagnose new diseases or infections, and refine their skills through such simulations [4]. Complex surgeries can be performed in such virtual environments which would reduce the risk of failure of any surgical

DOI: 10.1201/9781003449256-8

procedure on real patients. Such technology reduces risks, enhances training procedures, and it also accelerates the skill development of surgeons [5]. Therapy sessions can be provided with the help of virtual reality (VR) by using the concept of Metaverse that offers new ways to treat physical or mental health conditions. In order to improve the healthcare sector, Metaverse can be used to create an interactive and innovative educational platform that would help patients understand their medical conditions, methods of treatment, and prevention. It would help patients explore and understand their disease with the help of visuals through VR, which would enhance their health literacy [6]. Metaverse can be used to understand and visualize complex medical data which would help in better understanding of the current research and trends in the healthcare industry. VR in the Metaverse can be used for cognitive and neurological rehabilitation or treatment for individuals. The platform of Metaverse would be a great method to host virtual support groups which would enable people to connect with each other facing similar challenges. This would be a great approach for guidance or emotional support. A hospital might contain fewer nurses or working assistants. Due to such an issue, patients would not receive proper treatment or service from the hospital. The Metaverse offers us a potential solution with the help of using extended reality tools. Remote services can be provided by healthcare workers with the usage of augmented reality (AR) devices.

Game theory is a mathematical concept that helps to predict behavior. According to this theory, patients and doctors are considered players and they have a competing nature or agenda. In the field of healthcare, decisions play an important role in this sector. Game theory can help in providing the optimal solution and better decision-making models [7]. There might be a conflict of interest sometimes between doctors, patients, insurers, and regulators. Game theory helps in providing some way or method to a model and analyzes such interactions or situations, which helps in a better understanding of behavior and potential outcomes. This theory helps to analyze strategic situations. It can be effective to model the interactions between patients and the providers of healthcare such as doctors and hospitals. The patient has the option to choose between the treatment options which were suggested to him. Major factors such as cost, quality, and convenience can play a role in such options. The patient then decides and analyzes these factors to choose the treatment. Providers are the ones responsible for setting prices according to the demand for their services. In the recent era, healthcare insurance is very important for the majority of human beings. Game theory can help in providing insights and knowledge into the interactions between health insurance companies and the person who needs it. A person chooses health insurance on the basis of risk preferences and expenses, while insurance companies might try their best to design plans and insurance in such a way that can maximize the profit for them.

Game theory can be applied to study the virtual economies and markets within the Metaverse. Just as in real-world economies, individuals within the Metaverse make decisions about buying, selling, and trading virtual goods and services. Understanding these interactions can lead to insights into pricing strategies, resource allocation, and the emergence of virtual currencies.

As users interact with virtual spaces, virtual objects, and digital assets, questions of ownership and property rights become important [8]. Game theory can help design mechanisms for establishing and enforcing property rights within the Metaverse, considering factors like user incentives and potential conflicts. The Metaverse is built on social interactions, with users collaborating, competing, and forming relationships. Game theory can be used to model these interactions, predict behavior in social settings, analyze the formation of virtual communities, and design mechanisms to manage conflicts. Given the decentralized and complex nature of the Metaverse, governance and regulation mechanisms are crucial. Game theory can help design systems that incentivize positive behaviors, prevent abuse, and establish rules that promote fairness and cooperation.

Virtual spaces might have limited resources or spaces with high demand. Game theory can be applied to address questions of resource allocation, determining who gets access to what, and how competition for scarce resources is managed. Many aspects of the Metaverse are likely to involve decentralized applications (DApps) built on blockchain technology [9]. Game theory can assist in designing incentive mechanisms for DApps, encouraging participation, and maintaining security. With extensive personal data potentially being exchanged within the Metaverse, concerns about privacy and security arise. Game theory can help model potential security breaches, assess the effectiveness of privacy-preserving technologies, and analyze strategies for protecting users' information.

In Metaverse scenarios, collective decision-making may be required, such as regime change or local legislation. Game theory can help design voting mechanisms that encourage honest participation and prevent change. Game theory can inform systems for establishing and verifying digital identities and names in the Metaverse. This is critical for building trust among users to enable safe practices. Issues of intellectual property, copyright, and plagiarism can arise in the Metaverse where the materials are common. Game theory can help develop strategies that encourage creativity and respect the rights of creators. As VR becomes increasingly integrated into the Metaverse, collaborative environments can benefit from exploring game theory to optimize interaction patterns, shared resources, and collaborative practices.

Overall, game theory provides a powerful tool for understanding and shaping interactions, actions, and dynamics within the Metaverse. By applying game theory principles, developers, programmers, and stakeholders can create more engaging, just, and productive Metaverse experiences.

8.2 BACKGROUND

A revolutionary convergence is emerging in the rapidly evolving healthcare landscape that combines the participatory capabilities of the Metaverse with game theories of the process. The Metaverse, a virtual realm where digital and physical worlds meet, has captured the imagination of technology enthusiasts and researchers. Accordingly, game theory, a mathematical framework for understanding decision-making, has been applied across disciplines to optimize outcomes in complex situations.

Healthcare stands as a domain that is constantly seeking innovative solutions to address the complex challenges it faces. The healthcare industry is poised to change with the rise of cutting-edge technologies capable of transforming patient care,

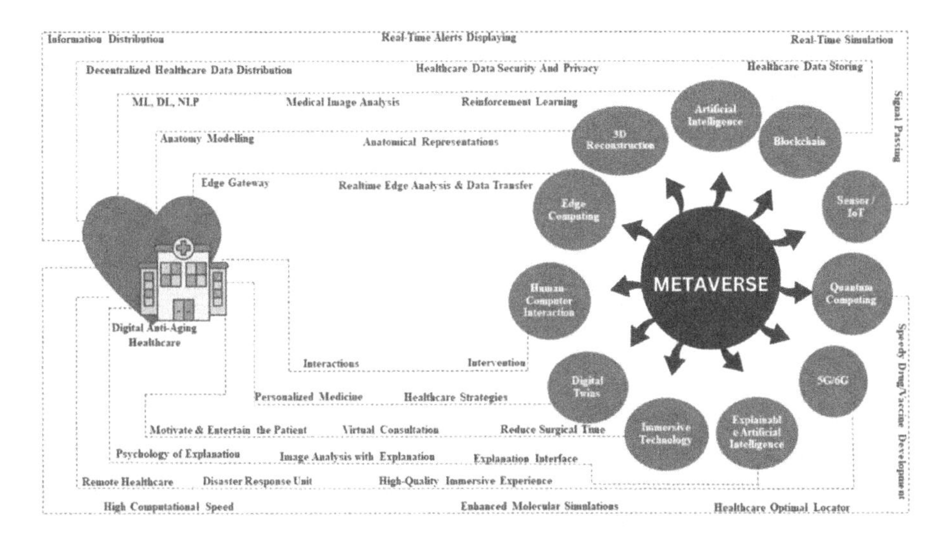

FIGURE 8.1 Implementation of Metaverse in the field of healthcare.

clinical practice, infrastructure, and medical research. In this context, the Metaverse emerges as an interesting way to integrate alternative approaches based on game theory.

The Metaverse presents an unprecedented opportunity to transform healthcare with game theory principles. This integration holds the promise of transforming patient care, treatment management, resource management, and medical research. The Metaverse is a canvas where healthcare professionals, patients, and researchers can interact in dynamic virtual environments, simulate real-world situations, and use a methodological approach derived from game theory. This session presents a new approach that challenges traditional health models.

Figure 8.1 specifies the role and importance of each major factor in Metaverse and the involvement of these factors in the healthcare industry.

8.2.1 Game Theory and Its Healthcare Applications

Central to this innovative synergy is the utilization of game theory in healthcare settings. Game theory, at its core, focuses on how rational individuals make choices in situations that involve interdependent decisions. When applied to healthcare within the Metaverse, this translates to collaborative and informed decision-making that considers multiple stakeholders' interests. This interdisciplinary blend harnesses the power of game theory to optimize resource allocation, enhance patient engagement, and refine clinical strategies, while the Metaverse provides the space for these theories to unfold practically.

Research has begun to delve into this exciting fusion of ideas. Scholars have demonstrated how game theory can be applied within the Metaverse to improve patient care, decision-making, and resource allocation. Studies indicate that the collaboration between virtual worlds and game theory can pave the way for innovative solutions in healthcare. Yazdanparast and Sepehri's work showcases the applicability of game theory principles in healthcare, illustrating its potential to enhance patient care dynamics. Similarly, Baniasadi and Masoodian's exploration of VR and game theory underscores the dynamic potential of this integration. Further insights from Winter and García-Jurado highlight how game theory's applications can be harnessed to improve healthcare resource allocation, a challenge that aligns with the Metaverse's potential to transcend traditional boundaries.

8.2.2 The Metaverse: Transforming Healthcare Delivery

While scholars have recognized the potential synergy between the Metaverse, game theory, and healthcare, the practical implications of this symbiotic convergence are only beginning to surface. At its core, the Metaverse represents a digital realm that transcends the limitations of physical presence. Its allure lies in the immersive experiences it offers-digital landscapes that mimic reality, enabling interactions that traverse geographical and temporal boundaries.

Moreover, as the Metaverse evolves, the boundary between reality and virtuality becomes increasingly blurred. Patients can participate in healthcare encounters from the comfort of their homes, and medical professionals can collaborate in virtual environments that foster knowledge exchange and real-time decision-making. The Metaverse enriches the healthcare experience, offering an avenue for more dynamic, engaging, and collaborative healthcare delivery.

Exploring Healthcare through the Metaverse: Delving deeper, several avenues emerge through which the Metaverse can be explored to reshape healthcare.

8.2.3 Virtual Clinical Simulations: Enriching Medical Expertise through Immersive Practice

In the rapidly evolving landscape of medical education and professional development, the integration of virtual clinical simulations within the Metaverse has garnered considerable attention [10]. The Metaverse, as a digital realm where the boundaries between real and virtual worlds blur, presents an ideal platform for medical professionals to engage in lifelike clinical scenarios [11]. These simulations offer a unique opportunity to hone clinical skills, refine decision-making processes, and enhance overall medical competence within a safe and controlled environment.

In a traditional healthcare setting, medical professionals often encounter a delicate balance between the desire to provide optimal patient care and the potential risks associated with untested interventions. The introduction of the Metaverse's virtual clinical simulations addresses this challenge by offering a risk-free zone for practice. Medical practitioners, from seasoned experts to trainees, can immerse themselves in complex medical scenarios, replicating real-world patient cases without exposing actual patients to any potential harm. This simulation environment provides a controlled space to experiment with a range of interventions, fostering a deeper understanding of the consequences of each decision made.

One of the paramount advantages of virtual clinical simulations is the ability to replicate diverse patient cases, spanning routine check-ups to rare medical emergencies.

This dynamic scope enables medical professionals to experience a broad spectrum of scenarios that they might encounter throughout their careers. For instance, a surgeon can practice intricate surgical procedures in a virtual operating room, adjusting their techniques and strategies based on the outcomes observed within the simulation. This iterative process of trial and refinement serves to not only bolster surgical skills but also to instill confidence in medical professionals.

Furthermore, the Metaverse's immersive nature enhances the authenticity of the simulation experience. Medical professionals can interact with virtual patients, medical equipment, and fellow healthcare team members, closely mirroring the complexity of real healthcare settings [4]. This level of realism promotes an emotional and cognitive connection, simulating the pressures and responsibilities that accompany real-life medical decision-making.

The benefits of virtual clinical simulations extend beyond individual skill development. They have the potential to foster collaborative learning and interdisciplinary cooperation. Medical teams can virtually convene to address complex cases, with each member contributing their expertise and insights. This collaborative approach mirrors the actual dynamics of a clinical team, where effective communication and cooperation are vital for successful patient outcomes.

As the Metaverse evolves and technology advances, the potential for even more sophisticated virtual clinical simulations becomes increasingly evident. The integration of haptic feedback, advanced physiological modeling, and machine learning algorithms could lead to simulations that mimic human responses and behaviors more accurately. This heightened realism would elevate the simulation experience to new heights, enabling medical professionals to explore the intricacies of medical decision-making in a virtual environment that closely resembles reality [12].

8.2.4 Patient-centered Virtual Care: Revolutionizing Healthcare Access and Engagement

Integrating affected person care into the virtual global is using alternatives in healthcare and affected person engagement. Metaverse breaks down the conventional boundaries of distance and physical barriers by means of performing as a bridge between patients and medical doctors. This pioneering method has the potential to revolutionize healthcare, mainly for populations dealing with challenges posed by far-flung areas and delivery issues.

Gaining access to fitness care in many components of the world may be a daunting task, in particular for human beings dwelling in sedentary or remote areas. Integrating virtual affected person care into Metaverse overcomes the same assignment with the aid of enabling affected person-medical doctor interaction from the consolation of their own home. With a digital session, sufferers can attend an appointment, get clinical recommendations, or even acquire an initial diagnosis without the arduous affected person journey.

This paradigm shift transcends mere accessibility and instead redefines the very experience of healthcare for patients. It eradicates the weight of long commutes to scientific centers, mainly while distance poses a giant impediment. Moreover, patients who require common medical interests or consultations stand to benefit from the ease of digital visits, trimming tour-related efforts, and time commitments, while nonetheless making sure of steady and regular interactions with their healthcare carriers.

The importance of patient-targeted virtual care becomes all of the greater reported because it breaks down geographical obstacles, allowing seamless connections between patients and clinical specialists regardless of their physical places. This inclusivity consists of particular weight for those situated in underserved regions or faraway areas wherein admission to comprehensive healthcare assets remains scarce. Digital care introduces a mechanism through which scientific know-how can permeate formerly marginalized zones, thereby addressing healthcare inequalities and selling identical right of entry to clinical services.

Another noteworthy factor of patient-focused virtual care lies in its capacity to expand affected persons' engagement and active participation in their healthcare adventure. The interactive nature of digital consultations empowers sufferers to collaborate in actual time with medical examiners, fostering discussions on remedy alternatives, addressing worries, and making knowledgeable selections tailored to their health needs. This collaborative approach shifts sufferers from passive recipients of care to lively partners, facilitating care plans that align with their particular circumstances.

However, the adoption of affected person-centered virtual care does usher in issues of its own. Safeguarding patient privacy and records security within the digital realm emerges as an important concern. Strong technological safeguards need to be hooked up to shield sensitive clinical facts and ensure adherence to stringent statistics safety rules. Moreover, healthcare practitioners have to acclimate to the nuances of undertaking virtual consultations, leveraging virtual tools effectively at the same time as keeping the personalized contact intrinsic to patient-centric care.

In summation, the infusion of patient-centered virtual care into the Metaverse presents a watershed advancement in healthcare accessibility and patient engagement. By transcending geographical barriers, cultivating equity in healthcare access, and empowering patients to actively participate in their wellness journey, this pioneering approach reshapes the contours of healthcare delivery. While challenges associated with data security and the integration of virtual care methodologies must be met, the potential benefits for patients—particularly those with mobility constraints or residing in remote areas—are boundless. As technological progress continues, patient-centered virtual care assumes an increasingly pivotal role in shaping the trajectory of modern healthcare [13].

8.2.5 Collaborative Medical Research in the Metaverse: Forging New Frontiers in Healthcare Advancement

Within the burgeoning Metaverse, the realm of collaborative medical research emerges as a promising avenue to revolutionize how healthcare insights are gained, shared, and translated into tangible advancements. As the Metaverse seamlessly converges into real and virtual spaces, it offers an innovative environment that transcends geographical limitations and disciplinary boundaries. This transformative potential presents a paradigm shift in the way researchers from diverse corners of the globe can collaborate, accelerating the pace of medical breakthroughs and reshaping the trajectory of healthcare progress [14].

Traditional medical research often unfolds within confined physical spaces and the scope of individual institutions or disciplines. The Metaverse dismantles these confines, enabling researchers to convene in shared virtual spaces unbounded by physical distance. This virtual interconnectedness invites experts from various fields—ranging from medicine and data science to engineering and genetics—to pool their expertise, insights, and resources. This collective intelligence enhances the depth and breadth of medical research, fostering a cross-disciplinary approach that can yield holistic and innovative solutions to complex healthcare challenges.

Moreover, collaborative medical research within the Metaverse offers a unique platform for global participation, leveling the playing field for researchers regardless of their location. This inclusivity enables voices from diverse backgrounds and regions to contribute to cutting-edge medical investigations. It not only fosters an equitable distribution of knowledge but also promotes the emergence of fresh perspectives and unconventional insights that can spark novel avenues of research.

The immersive and interactive nature of the Metaverse further elevates the collaborative research experience. Researchers can engage in virtual conferences, symposia, and workshops, transcending the limitations of time zones and travel constraints. The Metaverse's dynamic environment facilitates real-time interactions, discussions, and even the manipulation of virtual models and data sets. This degree of engagement enables more dynamic and engaging exchanges, paving the way for deeper collaboration and shared understanding among researchers.

However, this new frontier also raises questions and considerations that must be addressed. The Metaverse's virtual landscape demands a rethinking of how intellectual property, data ownership, and authorship are managed within collaborative research endeavors. Striking a balance between open collaboration and safeguarding individual contributions becomes crucial in maintaining ethical integrity and fostering sustained cooperation.

8.2.6 Optimizing Resource Allocation through Metaverse-driven Game Theory in Healthcare

Amidst the ever-evolving landscape of healthcare, the integration of Metaverse-driven game theory presents an innovative approach to tackle the intricate challenge of resource allocation. Game theory, a mathematical framework that dissects decision-making dynamics, finds a natural resonance within the Metaverse—a digital domain where real and virtual elements coalesce. This intersection opens up new avenues to optimize the distribution of scarce healthcare resources, harnessing the power of strategic thinking and simulations to foster efficient, equitable, and informed decision-making.

In the complex ecosystem of healthcare, resource allocation is a perpetual balancing act that demands careful consideration of limited resources against the diverse needs of patients and healthcare providers [15]. The integration of game theory within the Metaverse introduces a novel layer of analytical precision to this challenge. By employing strategic models and simulations, healthcare administrators can explore a spectrum of scenarios and potential outcomes before committing to a course of action. This process transcends traditional decision-making, fostering a more holistic and inclusive approach that considers multiple stakeholders and their varying priorities.

Game theory's application within the Metaverse also holds the promise of addressing the ethical quandaries inherent in resource allocation. Often, the distribution of resources involves competing interests and conflicting priorities [16]. Through Metaverse-driven simulations, administrators can visualize the impact of different allocation strategies on various groups, helping them navigate the intricate terrain of ethics and equity. This proactive approach not only improves the transparency of resource allocation decisions but also engenders a greater sense of trust among stakeholders, fostering a collaborative ethos in healthcare governance.

Furthermore, the Metaverse's immersive nature amplifies the impact of game theory-driven resource allocation strategies. Virtual simulations can replicate the intricate web of interactions that influence resource distribution, providing a dynamic platform to test and refine decision-making models. Administrators can experiment with various allocation scenarios, assessing the potential consequences of each strategy in a controlled and risk-free environment. This iterative process empowers administrators to make informed decisions that are not only strategically sound but also ethically defensible.

However, implementing Metaverse-driven game theory in healthcare resource allocation necessitates a robust technological infrastructure and a shift in organizational mindset. The convergence of intricate mathematical models, virtual simulations, and real-world healthcare data demands sophisticated computational capabilities. Additionally, stakeholders must be willing to embrace the Metaverse as a legitimate domain for strategic decision-making, necessitating a cultural shift toward embracing digital technologies and their transformative potential.

8.2.7 Gamified Health Education in the Metaverse: Nurturing Health Literacy through Interactive Engagement

The Metaverse's fusion with gamified fitness training represents a pioneering approach to reworking health literacy and selling knowledgeable decision-making among people. Health education, regularly hindered by way of records overload and disengagement, reveals a dynamic ally in the Metaverse—a realm wherein interactive reports captivate and educate concurrently. By means of infusing health education with gamification concepts within this digital domain, healthcare specialists have a completely unique possibility to revolutionize how individuals recognize and interact with their health.

Traditional fitness training methods regularly conflict to capture and preserve the eye of people bombarded with the aid of a steady inflow of information. Gamification, the combination of game elements into nonrecreation contexts, addresses this task head-on with the aid of introducing elements of undertaking, competition, and reward into academic reports. Inside the Metaverse, gamified health schooling becomes a dynamic tool to carry medical facts in an attractive and memorable manner.

Gamified fitness education's attraction lies in its potential to convert gaining knowledge from a passive level into a lively and immersive journey. Virtual environments in the Metaverse can simulate fitness scenarios, encouraging individuals to make choices, clear up challenges, and witness the outcomes of their alternatives. Through these interactive stories, complicated clinical principles are damaged down into relatable situations, empowering individuals to grasp and internalize crucial fitness records [17].

Furthermore, gamified health education nurtures a sense of empowerment and agency among individuals in managing their health. As they navigate through virtual challenges, individuals gain a deeper understanding of their health choices and the impact these choices have on their overall well-being. This experiential learning model extends beyond rote memorization, encouraging critical thinking and problem-solving skills that translate to informed health decisions in real life.

The Metaverse's immersive nature magnifies the impact of gamified health education [18]. Individuals can step into virtual healthcare environments, interact with avatars representing healthcare professionals, and partake in virtual health consultations. These interactions foster a sense of familiarity and comfort, reducing the anxiety often associated with medical encounters. Additionally, the Metaverse's interactive nature enables personalized learning pathways, tailoring educational experiences to an individual's unique health needs, preferences, and learning styles.

However, the successful integration of gamified health education within the Metaverse hinges on a delicate balance. Gamification should enhance learning rather than overshadow its educational intent. Ensuring that game elements serve as means to convey serious health information and promote well-being is paramount. Furthermore, the diversity of virtual learners' needs and preferences demands the creation of a wide array of engaging scenarios and game mechanics to cater to different learning profiles.

8.2.8 Ethical Dilemma Exploration in Virtual Environments: Fostering Critical Reflection in Healthcare

In the Metaverse's dynamic landscape, the integration of virtual environments for exploring complex ethical dilemmas stands as a powerful tool to cultivate ethical acumen among healthcare professionals. Ethics, a cornerstone of healthcare decision-making, often presents intricate scenarios where values, principles, and patient well-being intersect. Within the Metaverse's immersive settings, healthcare professionals have a unique opportunity to navigate these ethical intricacies, engage in critical reflection, and enhance their ability to make morally sound decisions in real-world healthcare contexts.

Traditional ethical education in healthcare often relies on theoretical instruction and case studies, which can lack the interactive and experiential dimensions needed to truly engage learners. The Metaverse's capacity to create interactive, immersive, and scenario-rich environments transforms ethical education into a tangible experience. Healthcare professionals can step into virtual healthcare scenarios, embody the roles of different stakeholders, and grapple with ethical dilemmas that mirror the complexities they encounter in their practice.

The Metaverse's immersive nature augments the depth of ethical exploration. Healthcare professionals can experience the emotional and psychological dimensions of ethical decision-making, gaining insight into the often nuanced and conflicting factors that shape such decisions. As they navigate virtual ethical scenarios, professionals are prompted to weigh different perspectives, anticipate consequences, and engage in introspective analysis. This process fosters empathy, moral sensitivity, and the ability to appreciate the multifaceted nature of ethical decisions.

Furthermore, the Metaverse's interactive potential enables collaborative ethical discussions among healthcare professionals across diverse specialties and cultural backgrounds. Virtual roundtable discussions can simulate interdisciplinary ethical consultations, where professionals contribute their insights, informed by their unique expertise and perspectives. This collaborative environment fosters a culture of open discourse, allowing participants to challenge assumptions, explore diverse viewpoints, and collectively arrive at ethically informed decisions.

However, the integration of ethical dilemma exploration in virtual environments necessitates careful design and consideration. Ensuring the authenticity and complexity of ethical scenarios is crucial for effective learning outcomes. Additionally, virtual ethical discussions must encourage constructive engagement while safeguarding the emotional well-being of participants as they grapple with challenging and emotionally charged scenarios.

8.2.9 Overcoming Challenges and Nurturing Readiness for Metaverse-driven Healthcare Innovation

While the integration of the Metaverse with healthcare brings forth a myriad of opportunities, it also demands a proactive approach to surmount challenges and equip healthcare professionals for this new frontier. The Metaverse's potential to reshape healthcare encounters, decision-making, and education hinges on overcoming technical, ethical, and cultural hurdles, while simultaneously fostering a mindset prepared to embrace innovation and adapt to new paradigms.

8.2.9.1 Technical Infrastructure and Accessibility

One of the primary challenges lies in building a robust technical infrastructure that can support the Metaverse's immersive experiences. This requires a balance between providing seamless virtual encounters and safeguarding the privacy and security of patient data. Additionally, ensuring universal accessibility is paramount, as not all individuals may possess the necessary technology or skills to navigate virtual environments. To tackle these challenges, investment in high-quality virtual platforms, ensuring data encryption, and offering training programs to enhance digital literacy become imperative.

8.2.9.2 Ethical Considerations and Patient Privacy

The virtual realm introduces novel ethical considerations that demand careful attention. Protecting patient privacy, securing virtual interactions, and obtaining informed consent within virtual healthcare encounters are pressing concerns. Ethical guidelines and frameworks must be established to ensure that Metaverse-driven healthcare practices maintain the same level of ethical integrity as traditional healthcare settings. Incorporating ethical education and training into healthcare professionals' curriculum is essential to ensure their readiness to navigate these new ethical complexities.

8.2.9.3 Cultural Transition and Mindset Shift

The adoption of Metaverse-driven healthcare practices also necessitates a cultural shift within the healthcare community. Healthcare professionals must be open to embracing digital technologies, adapting their practice to virtual environments, and leveraging innovative methodologies. Training programs that familiarize healthcare professionals with the Metaverse's potential, technical intricacies, and new approaches to patient care and decision-making are indispensable in nurturing this cultural transition.

8.2.9.4 Interdisciplinary Collaboration

The seamless convergence of Metaverse-driven healthcare requires collaboration among professionals from diverse disciplines, including medical experts, data scientists, VR specialists, and ethicists. Overcoming disciplinary boundaries and fostering a collaborative spirit are essential for the successful integration of the Metaverse within healthcare. Training programs that facilitate interdisciplinary understanding, communication, and cooperation are key to realizing the full potential of this convergence.

8.2.9.5 Patient-Centered Approaches

As virtual care becomes more prevalent, maintaining the human touch and patient-centeredness of healthcare encounters is paramount. Ensuring that patients feel heard, understood, and valued in virtual interactions requires training healthcare professionals to communicate effectively in a digital environment while retaining empathy and compassion.

8.3 METAVERSE USING DIFFERENT DOMAIN

8.3.1 Metaverse Integration across Diverse Domains: A Multifaceted Revolution

The emergence of the Metaverse as a transformative digital phenomenon has sparked a revolutionary wave that extends far beyond its impact on the healthcare sector. While its influence on healthcare is profound, the Metaverse's versatility transcends boundaries, permeating diverse domains such as education, entertainment, commerce, and social interaction. This evolution marks a profound shift in how human activities and interactions are mediated through technology. As we delve into the multifaceted dimensions of the Metaverse's integration, a panorama of possibilities unfolds, each domain echoing the Metaverse's potential to reshape industries, experiences, and human connections.

8.3.2 The Growth Trajectory

According to a comprehensive report by McKinsey, the global Metaverse market is projected to achieve an astonishing value ranging between $800 billion to $1 trillion by 2024. This exponential growth trajectory is a testament to the Metaverse's multi-faceted appeal and its capacity to redefine the way individuals engage with technology across various sectors.

8.3.3 Education

In the realm of education, the Metaverse has ushered in a paradigm shift that challenges traditional pedagogical approaches. Educators and institutions are harnessing the Metaverse's immersive capabilities to craft engaging learning environments. Platforms like Minecraft have become educational tools, offering students the opportunity to collaboratively construct and explore virtual worlds. This approach nurtures creativity, critical thinking, and problem-solving skills while breaking down the monotony of traditional classrooms. An example of this shift is Microsoft's "Minecraft: Education Edition," which hosts virtual classrooms that transport students to historical events or scientific landscapes. These interactive journeys foster an immersive learning experience that can make complex subjects more approachable and engaging.

8.3.4 Commerce

Commerce, too, has experienced a transformative upheaval with the integration of the Metaverse. Businesses are leveraging the Metaverse's immersive potential to redefine the shopping experience. Virtual stores are becoming a reality, enabling customers to browse and purchase products in interactive digital environments. Luxury brands are at the forefront of this innovation, embracing the Metaverse to host virtual fashion shows that meld high-end couture with digital artistry. Gucci's collaboration with "Roblox" to unveil its virtual Metaverse exhibition and Louis Vuitton's exploration of non-fungible tokens for exclusive digital fashion attest to the Metaverse's potential to redefine the very essence of consumer engagement.

8.3.5 Entertainment and Gaming

The entertainment industry, notably the gaming sector, has been a vanguard of the Metaverse revolution. VR and AR technologies have been seamlessly woven into gaming experiences, offering players a more immersive and interactive journey. Titles like "Fortnite" have gone beyond conventional gaming, transforming into digital Metaverse platforms. The Metaverse's impact on gaming is vividly illustrated by "Fortnite" virtual concerts, with artists like Travis Scott and Marshmello hosting groundbreaking digital performances that attracted millions of participants. These events epitomize the Metaverse's potential to transform entertainment into a dynamic shared experience that transcends geographical limitations.

8.3.6 Social Interaction and Work

In an era where remote work and virtual socialization have become the norm, the Metaverse is poised to revolutionize how we connect and collaborate. Platforms like Facebook's Horizon Workrooms are reimagining remote work by creating

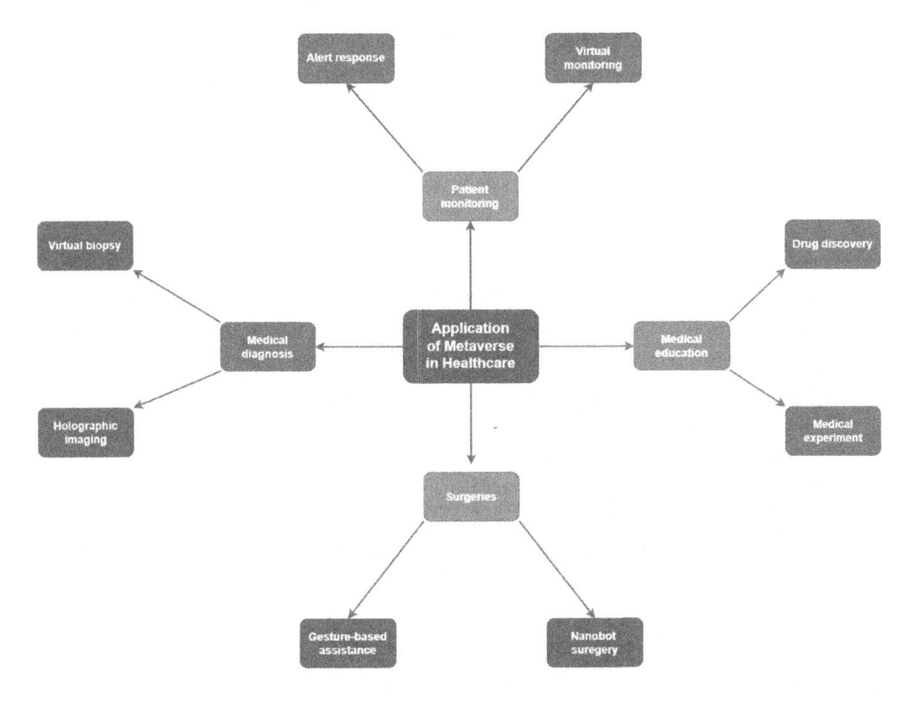

FIGURE 8.2 Connecting the healthcare sector with Metaverse. It also specifies how Metaverse can be used and implemented to transform the healthcare industry.

collaborative virtual spaces that emulate physical offices. This virtual workspace seeks to mitigate the isolation often associated with remote work, offering teams a sense of presence and proximity even in digital environments. Moreover, social VR platforms facilitate interaction beyond traditional social media, fostering meaningful connections and shared experiences in virtual realms [19].

In conclusion, the Metaverse's integration across diverse domains (shown in Figure 8.2) signifies a monumental shift in how humans interact with technology and with each other. Its versatility extends its influence beyond the confines of healthcare to reshape education, commerce, entertainment, and social dynamics. As projected market values soar and its applications continue to evolve, the Metaverse presents a dynamic and transformative landscape that transcends traditional boundaries. The multifaceted nature of its impact reaffirms the Metaverse's position as a digital phenomenon poised to redefine the future of human experiences and interactions [20].

8.4 METAVERSE IMPLEMENTATION PHASE

8.4.1 Implementation of Metaverse in Healthcare Using Game Theory: A Fusion of Innovation

In the realm of healthcare, the combination of the Metaverse, complemented by way of the concepts of game ideas, is sparking a seismic shift that has the capability to reshape the way healthcare is added, decisions are made, and resources are allotted.

This convergence of technologies and strategic selection-making frameworks introduces a new frontier in healthcare innovation. This paradigm isn't the handiest conceptually captivating however is likewise supported by means of an array of studies, research, insightful analyses, and compelling statistics that underscore the transformative influence of this dynamic fusion on the healthcare panorama.

8.4.2 Facts and Figures

1. **Global virtual reality market growth:** The global VR market, projected to reach a staggering valuation of $62.1 billion by 2027, has been witnessing exponential growth across various sectors, including healthcare. This growth trajectory is indicative of the diverse applications of VR, especially within the Metaverse context, and its potential to revolutionize healthcare interactions and experiences.
2. **Metaverse-driven medical training:** A groundbreaking study conducted by the National Center for Biotechnology Information unveiled the effectiveness of Metaverse-driven medical training. Medical students exposed to VR-based simulations within the Metaverse showcased remarkable improvements in their procedural skills and overall medical knowledge retention.
3. **Game theory for efficient resource allocation:** A pivotal application of game theory within the Metaverse is optimizing resource allocation in healthcare. A compelling study published in the *Healthcare* journal demonstrated how game theory-based models effectively facilitated the allocation of healthcare resources while considering intricate factors such as patient needs, costs, and ethical considerations.
4. **Telehealth and the Metaverse:** Amid the COVID-19 pandemic, telehealth usage experienced an astounding 1000% surge. The integration of the Metaverse in telehealth has further enhanced this trend by offering immersive virtual patient consultations, expanding healthcare accessibility, and transcending geographical limitations.

8.4.3 Strategic Decision-making Using Game Theory

1. **Efficient distribution:** The application of game theory principles to distribution ensures the efficient distribution of critical resources such as hospital beds, medical equipment, and personnel. This ensures that the classification process is fair, and efficient and reflects the needs of different stakeholders.
2. **Game theory in medicine:** Pharmaceutical companies use game theory models to design clinical trials that effectively balance factors such as sample size, trial duration, and patient selection, and inform drug design sensitive varieties and readily available drugs.
3. **Health insurance creation:** Metaverse-integrated game theory facilitates the design of health insurance policies that balance the insurer's profits with the goal of providing patients with accessible and affordable healthcare.
4. **Informed healthcare policies:** By employing game theory models, policymakers gain valuable insights into the behavior of different participants

within the healthcare ecosystem. This strategic insight guides the formulation of effective healthcare policies that align with the interests of various stakeholders.

The intersection of game theory and the Metaverse offers a dynamic approach to healthcare innovation. With its ability to optimize resource allocation, enhance strategic decision-making, and foster interdisciplinary collaboration, this convergence is poised to redefine healthcare in ways previously deemed unattainable. As the Metaverse continues to evolve and weave itself into the fabric of healthcare, it propels a new era characterized by innovation, inclusivity, and elevated patient outcomes. The integration of the Metaverse within healthcare's landscape serves as a testament to the powerful synergy between technology and strategic thinking, demonstrating how innovation can be harnessed to enhance human well-being and transform the healthcare experience.

8.4.4 Harnessing the Metaverse for Healthcare Transformation: A Multifaceted Beneficial Approach

The Metaverse, a dynamic digital realm where virtual and real elements converge, has emerged as a groundbreaking platform with immense potential to revolutionize the healthcare industry. This paradigm shift is driven by the integration of the Metaverse's immersive experiences with healthcare practices, resulting in multifaceted benefits that transcend traditional healthcare boundaries. As we delve into the transformative potential of the Metaverse in healthcare, a multitude of scenarios come to light, highlighting instances where the application of this technology could have mitigated challenges and improved patient care significantly.

8.4.5 Benefits of the Metaverse in Healthcare

1. **Enhanced medical training:** The Metaverse offers healthcare professionals immersive training experiences, allowing them to practice complex procedures in a risk-free virtual environment. Surgeons, for instance, can rehearse intricate surgeries repeatedly before performing them on real patients, resulting in increased precision and reduced medical errors.
2. **Patient-centered virtual care:** The Metaverse facilitates remote healthcare consultations, making healthcare accessible to individuals who face geographical or mobility barriers. Patients can virtually interact with healthcare providers, improving convenience and minimizing travel-related challenges.
3. **Gamified health education:** Leveraging the gamification potential of the Metaverse, healthcare education becomes engaging and memorable. Complex medical concepts can be broken down into interactive scenarios, enhancing health literacy and encouraging proactive health management.
4. **Resource allocation optimization:** The integration of game theory within the Metaverse enables healthcare administrators to simulate different resource allocation scenarios, ensuring efficient distribution of limited resources while considering various stakeholder interests.

5. **Ethical dilemma exploration:** Healthcare professionals can navigate complex ethical scenarios within the Metaverse, fostering critical reflection and enhancing their ethical decision-making skills, a critical aspect of patient care.
6. **Patient rehabilitation and therapy:** The Metaverse offers immersive virtual environments for patient rehabilitation and therapy. Physical therapy sessions can be transformed into engaging experiences that motivate patients to actively participate in their recovery.
7. **Global collaborative research:** Researchers from different parts of the world can collaboratively explore virtual medical simulations, contributing to global healthcare advancements. This collaborative approach accelerates the discovery of innovative treatments and diagnostic methods.
8. **Intellectual fitness interventions:** The Metaverse provides a secure space for individuals to interact in virtual remedy sessions for anxiety, phobias, and submit-annoying stress ailments. This approach complements accessibility and reduces stigma related to looking for an intellectual health guide.
9. **Chronic disorder management:** Digital assist businesses in the Metaverse can provide individuals with chronic situations a platform to share studies, search for advice, and obtain emotional guidance from friends going through similar challenges.
10. **Medical research simulation:** Researchers can simulate clinical experiments and clinical trials in the Metaverse, lowering the need for animal testing and expediting the invention of new medical breakthroughs.

8.4.6 Incidents that Could Have Been Handled with Metaverse

1. **Pandemic preparedness:** The Metaverse could facilitate global collaborative research and simulation of disease spread, aiding in preparedness for pandemics like COVID-19 [21].
2. **Medical training shortcomings:** Instances of medical errors due to inadequate training could be mitigated by Metaverse-driven immersive training experiences.
3. **Geographical barriers in access to specialists:** Patients in remote areas lacking specialist care could access virtual consultations through the Metaverse.
4. **Resource allocation during crises:** The integration of game theory within the Metaverse could optimize resource allocation during healthcare crises, ensuring efficient distribution [22].

The potential of the Metaverse in healthcare is vast, encompassing realms from medical education to patient care, resource optimization, and ethical dilemmas. By leveraging immersive experiences and strategic decision-making frameworks like game theory, the healthcare industry is poised for a transformative journey that enhances patient outcomes, empowers healthcare professionals, and contributes to the advancement of medical science [10, 20–23].

8.5 CONCLUSION: METAVERSE'S VITAL ROLE IN HEALTHCARE

The marriage of the Metaverse, combining VR, AR, and game theory, marks a paradigm shift in healthcare. This combination of technological wonder and methodological expertise holds the key to reshaping medical education, patient care, and resource management. In the 1960s, publishers like Evan Sutherland attempted a journey into the Metaverse by introducing computer-generated environments. In the decades that followed, VR and AR developed incrementally, with each leap marked by improvements every five years. Early studies focused on patient-centered modules, which empowered individuals to interactively understand complex medical concepts. The healing power of the Metaverse emerged, offering new ways to manage and treat pain. As game theory principles became mainstream, Metaverse became a formal companion. This improved the distribution of resources and transformed healthcare. Medical training has been revolutionized with VR simulations, providing a safe environment for complex physiotherapy exercises. The impact of the Metaverse transcended borders, facilitating global medical collaboration. The end of 2010 saw the end of the Metaverse. Virtual formal meetings followed the distance, as telemedicine evolved to meet the need for remote healthcare. Virtual solutions to ethical dilemmas were found, enhancing the decision-making skills of professionals. Real-life footage and data confirmed the Metaverse's potential. From pandemic preparedness to complex surgical simulations, there have been situations where the Metaverse can provide highly effective solutions. In conclusion, the integration of Metaverse into healthcare demonstrates the convergence of technology and strategy. Its evolution from conception to transition highlights its potential to redefine health practice. As Metaverse continues to evolve, it promises innovation, a better patient experience, and a future where reality and virtuality seamlessly merge to improve healthcare.

REFERENCES

1. C. T. Nguyen, D. T. Hoang, D. N. Nguyen and E. Dutkiewicz, "Meta-chain: A novel blockchain-based framework for Metaverse applications," in 2022 IEEE 95th Vehicular Technology Conference: (VTC2022-Spring), Helsinki, Finland, 2022, pp. 1–5. doi: 10.1109/VTC2022-Spring54318.2022.9860983.
2. A. Athar, S. M. Ali, M. A. I. Mozumder, S. Ali and H.-C. Kim, "Applications and possible challenges of healthcare Metaverse," in 2023 25th International Conference on Advanced Communication Technology (ICACT), Pyeongchang, Korea, Republic of, 2023, pp. 328–332. doi: 10.23919/ICACT56868.2023.10079314.
3. J. M. Ruiz Mejia and D. B. Rawat, "Recent advances in a medical domain Metaverse: Status, challenges, and perspective," in 2022 Thirteenth International Conference on Ubiquitous and Future Networks (ICUFN), Barcelona, Spain, 2022, pp. 357–362. doi: 10.1109/ICUFN55119.2022.9829645.
4. H. F. Mahdi, B. Sharma, R. Sille and T. Choudhury, "Metaverse and healthcare: An empirical investigation," in 2023 5th International Congress on Human–Computer Interaction, Optimization and Robotic Applications (HORA), Istanbul, Turkiye, 2023, pp. 1–6. doi: 10.1109/HORA58378.2023.10156688.
5. A. S. Bajkoti, S. Tiwari, T. Ranjan, R. S. Bisht and A. Mittal, "Role of quantum computing and Metaverse in the field of healthcare and medicine," in 2023 5th International Conference on Inventive Research in Computing Applications (ICIRCA), Coimbatore, India, 2023, pp. 306–311. doi: 10.1109/ICIRCA57980.2023.10220921.

6. R. Jeljeli, F. Farhi, F. Shawabkeh, A. A. Al Marei, M. M. Mohammed and S. Setoutah, "The role of self-determination theory in adopting Metaverse for healthcare and diagnostics among healthcare professionals," in 2023 International Conference on Multimedia Computing, Networking and Applications (MCNA), Valencia, Spain, 2023, pp. 95–102. doi: 10.1109/MCNA59361.2023.10185832.
7. M. Wooldridge, "Does game theory work?," *IEEE Intelligent Systems*, vol. 27, no. 6, pp. 76–80, 2012. doi: 10.1109/MIS.2012.108.
8. M. Cruz, A. Oliveira and A. Pinheiro, "Flowing through virtual animated worlds—Perceptions of the Metaverse," in 2022 Euro-Asia Conference on Frontiers of Computer Science and Information Technology (FCSIT), Beijing, China, 2022, pp. 241–245. doi: 10.1109/FCSIT57414.2022.00057.
9. J. Moodley, et al., "Beyond reality: An application of extended reality and blockchain in the Metaverse," in 2023 IEEE International Conference on Omni-layer Intelligent Systems (COINS), Berlin, Germany, 2023, pp. 1–4. doi: 10.1109/COINS57856.2023.10189285.
10. A. Sihna, H. Raj, R. Das, A. Bandyopadhyay, S. Swain and S. Chakrborty, "Medical education system based on Metaverse platform: A game theoretic approach," in 2023 4th International Conference on Intelligent Engineering and Management (ICIEM), IEEE Conference, 2023, pp. 1–6.
11. R. Chengoden, et al., "Metaverse for healthcare: A survey on potential applications, challenges and future directions," *IEEE Access*, vol. 11, pp. 12765–12795, 2023. doi: 10.1109/ACCESS.2023.3241628.
12. R. Papara, R. Galatus and L. Buzura, "Virtual reality as cost effective tool for distance healthcare," in 2020 22nd International Conference on Transparent Optical Networks (ICTON), Bari, Italy, 2020, pp. 1–6. doi: 10.1109/ICTON51198.2020.9203420.
13. M. F. Li and J. Feng, "Healthcare road map to modernization in Clouds: Healthcare forum for healthcare professionals, medical device manufacturers, pharmaceutical companies and average people on Virtual Private Clouds," in 2017 IEEE/ACM International Conference on Connected Health: Applications, Systems and Engineering Technologies (CHASE), Philadelphia, PA, USA, 2017, pp. 247–248. doi: 10.1109/CHASE.2017.86.
14. A. K. Bashir, N. Victor, S. Bhattacharya, T. Huynh-The, R. Chengoden, G. Yenduri, P.K.R. Maddikunta, Q.V. Pham, T.R. Gadekallu, and M. Liyanage, "Federated learning for the healthcare Metaverse: Concepts, applications, challenges, and future directions," *IEEE Internet of Things Journal* (2023). doi: 10.1109/JIOT.2023.3304790.
15. J. Puustjärvi and L. Puustjärvi, "Resource allocation in healthcare: Implications of scarce resources and temporal constraints," in 2013 IEEE International Conference on Industrial Engineering and Engineering Management, Bangkok, Thailand, 2013, pp. 1067–1071. doi: 10.1109/IEEM.2013.6962574.
16. W. Shao-Jen, T. Wu, G. Mackulak and J. Fowler, "Distributed resource allocation for healthcare systems," in 2008 IEEE International Conference on Service Operations and Logistics, and Informatics, Beijing, China, 2008, pp. 1078–1083. doi: 10.1109/SOLI.2008.4686559.
17. C. Sotirakou, S. Papavasiliou, C. Mourlas and K. Van Isacker, "Gamified mobile/online learning for personal care givers for people with disabilities and older people," in 2015 International Conference on Interactive Technologies and Games, Nottingham, UK, 2015, pp. 22–27. doi: 10.1109/iTAG.2015.16.
18. S. Balaji, V. Rajaram and S. Sethu, "Gamification of learning using 3D spaces in the Metaverse," in 2023 International Conference on Networking and Communications (ICNWC), Chennai, India, 2023, pp. 1–5. doi: 10.1109/ICNWC57852.2023.10127461.
19. N. He, K. Ding and J.-B. Zhang, "Exploration and research on digital education scenarios from the perspective of Metaverse," in 2022 10th International Conference on Orange Technology (ICOT), Shanghai, China, 2022, pp. 1–4. doi: 10.1109/ICOT56925.2022.10008167.

20. P. Gupta, K. Bhadani, A. Bandyopadhyay, D. Banik and S. Swain, "Impact of Metaverse in the near 'future'," in 2023 4th International Conference on Intelligent Engineering and Management (ICIEM), IEEE Conference, pp. 1–6, 2023.
21. Z. Lv, "Virtual reality based human–computer interaction system for Metaverse," in 2023 IEEE Conference on Virtual Reality and 3D User Interfaces Abstracts and Workshops (VRW), Shanghai, China, 2023, pp. 757–758. doi: 10.1109/VRW58643.2023.00221.
22. A. Bandyopadhyay, A. Sarkar, S. Swain, D. Banik, A. E. Hassanien, S. Mallik, A. Li and H. Qin, "A game-theoretic approach for rendering immersive experiences in the Metaverse," *Mathematics*, vol. 11, no. 6, p. 1268, 2023.
23. A. Pandey, A. Chirputkar and P. Ashok, "Metaverse: An innovative model for healthcare domain," in 2023 International Conference on Innovative Data Communication Technologies and Application (ICIDCA), Uttarakhand, India, 2023, pp. 334–337. doi: 10.1109/ICIDCA56705.2023.10099764.

9 Fake Medicine Detection Using Blockchain in Metaverse Domain

Anjan Bandyopadhyay
School of Computer Science and Engineering, Kalinga Institute of Industrial Technology, Bhubaneshwar, India

Jyotirmoy Karmakar
School of Computer Science and Engineering, Kalinga Institute of Industrial Technology, Bhubaneshwar, India

Kartikeya Raj
School of Computer Science and Engineering, Kalinga Institute of Industrial Technology, Bhubaneshwar, India

Sujata Swain
School of Computer Science and Engineering, Kalinga Institute of Industrial Technology, Bhubaneshwar, India

9.1 INTRODUCTION

The proliferation of counterfeit medicines has emerged as a critical global issue, posing profound threats to public health, healthcare systems, and patient trust. In the real world, counterfeit drugs infiltrate supply chains, undermining the efficacy of medical treatments and jeopardizing patient well-being [1]. According to a recent assessment by the World Health Organization [2], approximately 10.5% of pharmaceutical drugs found in markets across low and middle-income nations are identified as counterfeit. This underscores the pressing need for the development of a robust model aimed at tackling the pervasive issue of counterfeit drugs. These spurious pharmaceuticals not only result in treatment failures and worsened health outcomes but also contribute to the development of drug-resistant pathogens, thus exacerbating existing healthcare challenges. Beyond the confines of the physical realm, the emergence of the virtual Metaverse has introduced new dimensions to this problem. In this immersive digital environment, where interactions and transactions mimic real-life scenarios, the risk of counterfeit medicines infiltrating the virtual pharmaceutical landscape is an alarming prospect. The Metaverse, with its dynamic and borderless nature, offers both opportunities and vulnerabilities [31–33]. Addressing the convergence of counterfeit medicines within both the real world and the Metaverse is of paramount significance [3]. As technological advancements enable the interplay between reality

 DOI: 10.1201/9781003449256-9

and virtuality, an innovative solution is warranted to safeguard health and authenticity across these interconnected domains. In this context, the integration of BCT within the Metaverse presents a pioneering approach to combat the fake medicine dilemma, promising enhanced security, transparency, and accountability in pharmaceutical transactions within this emergent digital realm.

9.1.1 Blockchain Technology

BCT, originally conceptualized as the underlying framework for cryptocurrencies like Bitcoin, has evolved into a versatile and transformative innovation with far-reaching implications. At its core, blockchain is a decentralized and distributed digital ledger that securely records and verifies transactions in a transparent and tamper-proof manner [4]. Unlike traditional centralized databases, where a single entity maintains control, blockchain operates as a network of interconnected nodes that collectively validate and store data. This consensus-driven mechanism ensures immutability and integrity [5], making it an ideal solution for scenarios requiring secure and auditable record-keeping. Each block in the chain contains a cryptographic hash of the previous block, creating an interlinked and chronological sequence that prevents unauthorized modifications. This ground-breaking technology has transcended its financial origins and found application in various sectors, revolutionizing the way data are managed and shared [6].

9.1.2 Concerns with Conventional Pharmaceutical Industry

The traditional pharmaceutical industry has been plagued by a persistent and alarming issue of fake medicines, posing significant risks to global public health. This has been highlighted even further during the COVID-19 pandemic. The urgent global demand for treatments and vaccines created an environment ripe for exploitation by counterfeiters seeking to profit from the crisis [7]. This problem stems from various loopholes in the supply chain and regulatory systems, allowing counterfeit or substandard medications to infiltrate the market. Counterfeit drugs often lack the active ingredients necessary for therapeutic efficacy, potentially leading to treatment failures, drug resistance, and adverse health effects. These counterfeit products are adeptly marketed to resemble genuine medications, making them difficult for consumers to discern. Inadequate regulatory enforcement [8], particularly in certain regions, coupled with limited public awareness, exacerbates the problem. Efforts to address this crisis require international collaboration, robust supply chain security measures, stricter regulations, and improved public education to safeguard the integrity of pharmaceutical products and ensure the well-being of patients worldwide.

9.1.3 Blockchain in Healthcare

In the healthcare domain, BCT offers a promising avenue for addressing critical challenges related to data security, interoperability, and patient privacy. By enabling secure and transparent sharing of medical records, research data, and patient information, blockchain can streamline data exchange between disparate healthcare systems while maintaining strict control over data access [8]. Additionally, it can empower patients with greater ownership of their health data, allowing them to grant permission to specific entities for data usage. This enhanced data interoperability can lead to more accurate diagnoses, improved treatment plans, and expedited research collaborations. Furthermore, blockchain's potential extends into the Metaverse domain, where the fusion of digital experiences and economic transactions takes place [9]. The Metaverse, an immersive virtual reality that transcends geographical boundaries, can leverage blockchain to ensure the authenticity of digital assets, intellectual property rights, and virtual identities. Through decentralized ownership records and smart contracts (SCs), blockchain can establish trust and security in Metaverse interactions, enabling secure virtual transactions, authentic digital ownership, and the creation of a seamless and trusted digital ecosystem [11]. The convergence of BCT with healthcare and the Metaverse holds the promise of reshaping these domains, fostering innovation, and enhancing user experiences while preserving data integrity and user control.

9.1.4 Consensus Mechanisms

Consensus mechanisms form the bedrock of decentralized networks, orchestrating agreement among distributed nodes and bolstering the credibility of transactions, making them a pivotal element in BCT. These mechanisms play a pivotal role in achieving agreement among distributed nodes, ensuring the accuracy and integrity of transactions and data across the network [12]. By establishing a unified decision-making process, consensus mechanisms mitigate the risk of malicious attacks and maintain the consistency of the distributed ledger. These mechanisms are instrumental in countering Byzantine and Sybil attacks [13], two significant security challenges. Byzantine fault tolerance, achieved through consensus, safeguards against malicious actors attempting to disrupt the network by disseminating conflicting information. Consensus mechanisms impose stringent agreement criteria, thwarting attempts to manipulate the system. Additionally, the Sybil attack, wherein adversaries create multiple pseudonymous identities to gain undue influence, is mitigated by consensus protocols like proof of work (PoW) and proof of stake (PoS). PoW's computational complexity and PoS's requirement of holding a stake act as barriers to such attacks, reinforcing the network's robustness. The study of consensus mechanisms extends beyond transaction validation, encapsulating strategies that ensure the network's resilience against a spectrum of adversarial activities.

9.1.5 Smart Contracts

SCs represent integral components of self-executing code designed to trigger automatically upon the fulfillment of predetermined conditions. Their emergence stems from the integration of BCT across diverse sectors, addressing the need for streamlined incorporation of business logic. These SCs exhibit the capability to not only autonomously execute specific functions but also to interact seamlessly with other SCs, facilitating the collaborative execution of complex business operations [14]. Within the context of developing decentralized applications (DApps), SCs offer numerous advantages. However, their utility is accompanied by inherent limitations, prominently including the inability to access external application programming interfaces and engage with real-world data. To surmount these constraints, the proposed architecture incorporates oracles, specialized components that enable SCs to bridge the gap between blockchain and external data sources. By interfacing SCs with oracles [15], the architecture augments the capacity of SCs to securely and reliably interact with off-chain information, thereby enhancing their potential to drive innovative and functional blockchain-based applications.

9.2 LITERATURE REVIEW

This section delves into noteworthy contributions involving the application of BCT within various domains of the medical business. The study by the authors of Ref. [16] exemplifies a comprehensive exploration of blockchain's potential across multiple facets of the medical industry, ranging from drug research to the management of health sectors. Notably, their work encompasses insightful strategies for obviating intermediaries within the medical landscape, redefining transactional paradigms. In parallel, Choo et al. [17] provide a forward-looking perspective on the future trajectory of blockchain's integration within the medical sector. This scholarly endeavor not only identifies unaddressed dimensions but also accentuates emergent criteria heretofore neglected in the blockchain discourse. Furthermore, this chapter delves into research intricacies, inclusive of scalability, the block withholding attack, and the nuanced landscape of blockchain mining incentives. The nuanced analysis extends to their applicability or potential absence within the healthcare domain. In essence, these investigations collectively contribute to the scholarly fabric, enriching our comprehension of blockchain's evolving role within the intricate realm of medical business. In the realm of medical research, Junejo et al. [18] undertook a comprehensive assessment of prior studies, with a specific focus on contemporary investigations marked by a decentralized approach, notably exemplified by BCT. Their work elucidates the applicability of this decentralized paradigm in the context of brain systems, exemplified through the successful preservation of a simulated digital brain within a decentralized network framework, such as blockchain. The authors notably elucidated pivotal challenges that currently hinder the seamless integration of blockchain within the medical industry. It is noteworthy that each blockchain-based model inherently operates through the conduit of SCs. Parallelly, Kumar et al. [19] proffered an analogous system tailored to the medical sector. Beyond their contributions, their study extensively outlined critical challenges that beset the application of blockchain in healthcare, encompassing facets like scalability limitations, escalated developmental expenses, intricate standardization dilemmas, cultural resistance, legal ambiguities, and more. In essence, these scholarly endeavors collectively illuminate the evolving landscape of blockchain in medical domains while intricately dissecting the hurdles that necessitate attention to harness its transformative potential effectively.

The work of Bell et al. [20] emerges as a critical contribution in the discourse of contemporary healthcare challenges. Their meticulous examination not only highlights the pressing intricacies faced by the healthcare profession but also adeptly discerns strategic domains where the integration of BCT holds transformative potential. By delineating contexts such as clinical trials, inter-organizational data sharing encompassing hospitals and manufacturers, and the intricate domain of patient records, the authors underscore blockchain's applicability as a problem-solving instrument within these pivotal sectors. An intriguing revelation within their study pertains to the hesitance exhibited by prevailing systems, particularly supply chains, which serves as the primary deterrent to the comprehensive adoption of blockchain within the healthcare spectrum. In a parallel vein, the scholarly endeavors of the authors of Ref. [21] cast an illuminating light upon blockchain's varied application scenarios. They not only dissect these cases but also meticulously analyze, evaluate, and present a coherent methodology that charts the integration of blockchain into the existing healthcare procedures. Collectively, these scholarly pursuits contribute substantively to the enriched understanding of blockchain's imminent potential in revolutionizing healthcare while concurrently mapping out the requisite pathways for its seamless assimilation. The study's findings illuminate the prevalent applications of BCT in healthcare, notably electronic health records (EHR) and personal health records, with Ethereum and Hyperledger Fabric emerging as dominant open-source frameworks for constructing blockchain-based applications. The authors of Ref. [22] emphasize the critical privacy concerns inherent in data warehousing and sharing platforms, adeptly mitigated through the adoption of a permissioned blockchain, thereby addressing vulnerabilities in data transfer and eliminating the potential single point of failure. Their creation of a smartphone application for data collection reinforces practical application. Similarly, Dwivedi et al. [23] underscore the paramount significance of privacy and security in data exchange and storage within the healthcare context, highlighting the pressing need for comprehensive safeguards. Collectively, these works provide foundational insights into blockchain's transformative potential in healthcare, intricately addressing challenges linked to data privacy and security.

The scholarly work conducted by the authors of Ref. [24] casts a discerning spotlight on the critical concern of medication safety, expertly endeavoring to ameliorate this challenge through the strategic integration of BCT coupled with QR codes. Within their comprehensive analysis, the authors adeptly elucidate the existing deficiencies within the pharmaceutical supply chain, pinpointing the anomalies that necessitate remediation. In response to this exigency, they present a meticulously crafted methodology, seamlessly incorporating a blockchain-based architecture into the supply chain framework. This recommended course of action stands as a testament to their dedication to ensuring not only the reliability of medications but also the unequivocal authenticity of the entities engaged in their production. In essence, their scholarly contribution enriches our understanding of how blockchain, synergistically aligned with QR codes, can invigorate medication safety, thereby making a noteworthy stride toward the enhancement of pharmaceutical quality and integrity.

In their scholarly inquiry, Hag et al. [25] identified inherent complexities within the extant pharmaceutical supply chain and emphasized the applicative potential of blockchain as an alternative framework. The authors advocated for the utilization of a permissioned blockchain to effectively store comprehensive network data, thereby ensuring both traceability and transparency in the movement of specific entities across different levels. This proposed approach not only addresses existing challenges but also guarantees the engagement of trustworthy participants within the network, highlighting the transformative role of BCT in enhancing pharmaceutical supply chain dynamics.

The authors in Ref. [26] underscored the medicinal implications inherent in a decentralized paradigm, exemplified by BCT. Their investigation encompassed a comprehensive exploration of blockchain's applicability across diverse domains, including EHR, medical insurance, biomedical field, and medical supply chain. Ultimately, their findings indicated that despite its multifaceted problem-solving capabilities, BCT still awaits widespread adoption within healthcare institutions. The authors assert that decision-makers in healthcare should possess a comprehensive understanding of the technology's potential and its transformative prowess, advocating for its seamless integration into the existing healthcare framework.

The research presented in Ref. [27] introduces an all-encompassing electronic health network utilizing BCT to effectively address India's issue with counterfeit pharmaceuticals. The proposed solution capitalizes on the inherent capabilities of blockchain networks to meticulously monitor the entire trajectory of medicinal supply, spanning from manufacturers to end users. In instances where spurious medications infiltrate the system at any point along this continuum, the blockchain-based mechanism promptly detects the anomaly and promptly halts its propagation. Through simulation on a Hyperledger Fabric platform, the system's performance is meticulously evaluated and benchmarked against alternative methodologies. The findings underscore that while the proposed system necessitates a computationally intensive approach, it inherently offers a dependable and trustworthy resolution to the counterfeit medication predicament.

9.3 PROPOSED METHODOLOGY

This section elaborates on the envisioned framework's specifics. The envisaged framework presents a comprehensive and cohesive solution designed to empower manufacturers, pharmacists, and Drug Regulatory Authorities (DRAs) with real-time insights for enhancing the management of drug distribution. Furthermore, this framework proposes the assignment of distinctive batch numbers to medications available at registered pharmacies, amplifying the credibility and dependability of the vendors. Leveraging BCT, transactions occurring between involved stakeholders will be meticulously recorded, ensuring undeniable authenticity. The inherent attributes of blockchain, including its decentralized, distributed, transparent, and immutable nature, can be strategically harnessed as a potent instrument for detecting fraudulent pharmaceuticals.

Illustrated in Figure 9.1, the envisioned framework embodies a network of diverse participants including manufacturers, pharmacists, DRA, and customers, all interlinked within a decentralized network structure. Each of these integral components within the supply chain assumes the role of a node on the public blockchain, signifying their distinct presence. Every node is accompanied by an individual account, meticulously representing its unique identity. Notably, the DRA is designated as the authoritative steward of the blockchain, vested with overarching control and

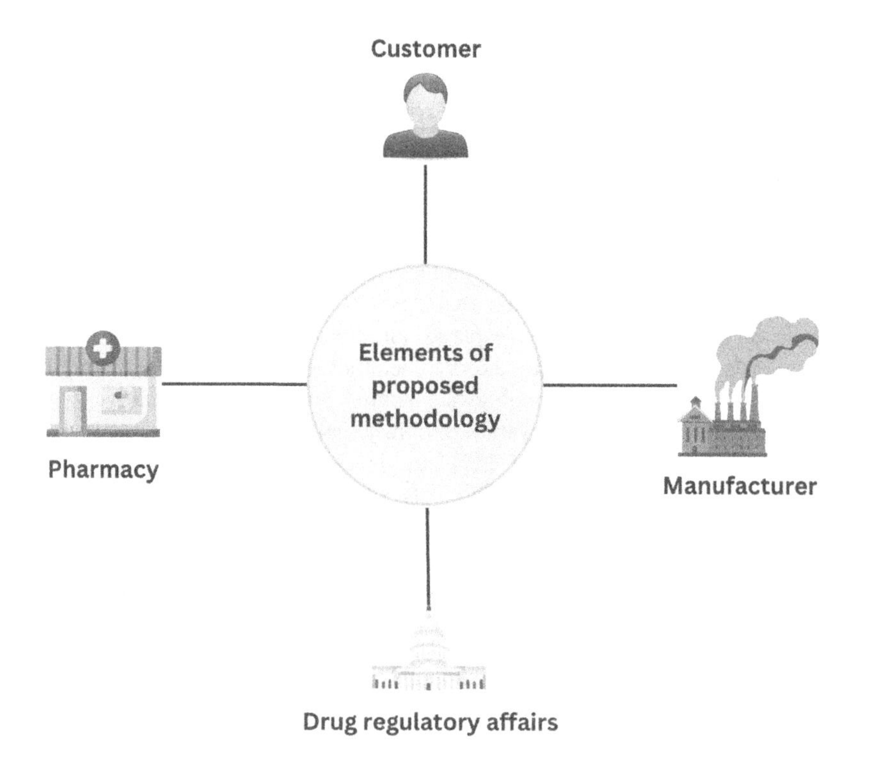

FIGURE 9.1 Elements of proposed methodology.

surveillance over the end-to-end drug distribution process, thereby ensuring unwavering security and the utmost quality of pharmaceutical products.

Within our outlined methodology, a meticulous provision has been made to affix QR codes onto individual batches of medicines. This strategic implementation serves a dual purpose: first, facilitating the seamless identification of the authentic producer for each medicine batch, and second, significantly elevating the traceability aspect within the system. For the underpinning blockchain infrastructure, we have judiciously opted for the polygon Matic blockchain. This deliberate choice is grounded in its notable attributes of swiftness and remarkably nominal gas fees. The adoption of the polygon Matic blockchain thus aligns seamlessly with our endeavor to establish an efficient and cost-effective ecosystem for our proposed solution.

9.3.1 Medicine Tokenization Process

The method presented outlines a tokenization process leveraging non-fungible tokens (NFTs) to guarantee the genuineness and traceability of drugs throughout the pharmaceutical supply chain. The integration of the ERC-721 standard, a widely recognized Ethereum token standard developed for NFTs, simplifies the implementation of this process. Within this framework, the application of NFTs offers a holistic solution to challenges such as counterfeit drugs and the authentication of legitimate origins.

There are various steps involved in the tokenization process. They are discussed below:

1. **Contract deployment and initialization:** The SC is deployed on the polygon Matic blockchain and is constructed using the ERC-721 standard. The ERC-721 SC is imported from the openzeppelin library, guaranteeing a robust security framework. This norm facilitates the creation of distinct and indivisible tokens, enabling the creation of distinct pharmaceutical batches.
2. **Medicine and pharmacy registration:** In the methodology we employ, authorized pharmacies granted the capability to input medicine-related details onto the blockchain. These medicinal items are encapsulated within NFTs, with each distinct NFT representing a particular batch of medicine. Prior to engaging with the contract, pharmacies are required to fulfill verification prerequisites, a measure implemented to guarantee the involvement of only credible sources.
3. **Verification and authentication:** The contract provides the functionality to authenticate both pharmacies and medicines. Pharmacies have the option to undergo a verification process to establish their authenticity and trustworthiness. Conversely, medicines are designated as validated and authorized once their genuineness has been verified.
4. **Non-fungible tokens creation:** Upon successful verification of a medicine, an NFT is created specifically for that batch. This NFT functions as a digital testament of legitimacy, offering an exclusive representation of the respective medicine. The act of minting produces a distinctive identifier (token ID) intrinsically linked to the NFT.
5. **Ownership and traceability:** Every individual NFT aligns with a distinct batch of medicine, thereby facilitating efficient monitoring and traceability across the supply chain. Possession of an NFT denotes ownership of the authorized medicine batch it symbolizes.
6. **Blockchain immutability and transparency:** The inherent immutability of the blockchain ensures that information pertaining to NFTs, encompassing medicine verification status and ownership, remains unmodifiable once it is documented. This assurance establishes an impervious record of the complete medicine lifecycle.

The suggested SC uses QR codes as a practical and user-friendly technique of verifying the legitimacy of pharmaceuticals in combination with the tokenization process utilizing NFTs. The integration improves the pharmaceutical supply chain's overall traceability and transparency by allowing stakeholders to quickly validate drug batches by scanning QR codes.

9.3.2 Smart Contract Architecture

A SC is a self-executing agreement encoded in computer code that automates, verifies, and enforces the terms and conditions of a contract. Operating on BCT, an SC eliminates the need for intermediaries, thereby streamlining and enhancing the

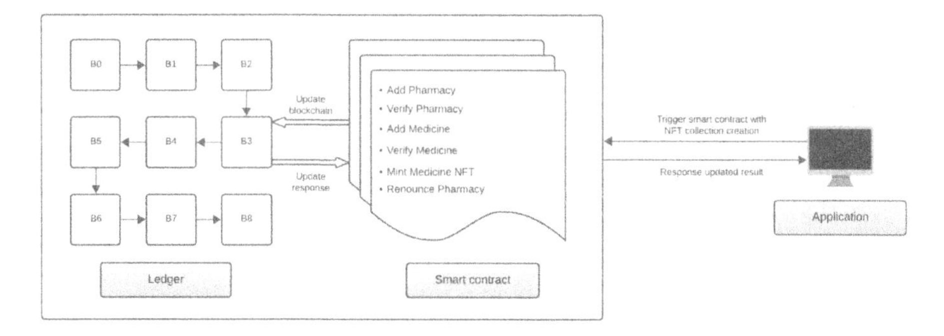

FIGURE 9.2 Smart contract architecture.

reliability of various transactions and agreements. It executes predefined actions when specific conditions are met, offering a secure, transparent, and decentralized framework for parties to engage in trustless interactions. This technological innovation has applications across diverse industries, from finance to supply chain management, revolutionizing the way agreements are formulated and executed in the digital era [28].

The architectural framework of our SC, as depicted in Figure 9.2, encompasses a range of essential features aimed at identifying and thwarting counterfeit medicines within the Metaverse. Subsequently validated and created, NFTs representing batches of authentic medicines are showcased in the Metaverse marketplace, providing prospective buyers the opportunity to purchase them using the Matic token. Using polygon Matic blockchain, we reduce transaction costs and enhance security and speed, given in Table 9.1.

The methodology is elucidated through Figure 9.3. In the initial steps, the DRA establishes the NFT collection, followed by the registration of pharmacies into this ecosystem. Subsequently, the DRA conducts the necessary verification of the registered pharmacies. Once verified, these pharmacies gain authorization to create and offer medicines within the Metaverse marketplace, given that the medicines

TABLE 9.1
Smart Contract Functions, Access Modifiers, and Event Generations

Smart contract functions	Modifiers	Event generated
addPharmacy	Any	None
verifyPharmacy	onlyOwner()	PharmacyVerified
addMedicine	Any	MedicineAdded
verifyMedicine	onlyOwner()	None
isMedicineVerified	Any	None
renouncePharmacy	onlyOwner()	PharmacyRevoked
Withdraw	onlyOwner()	Withdraw

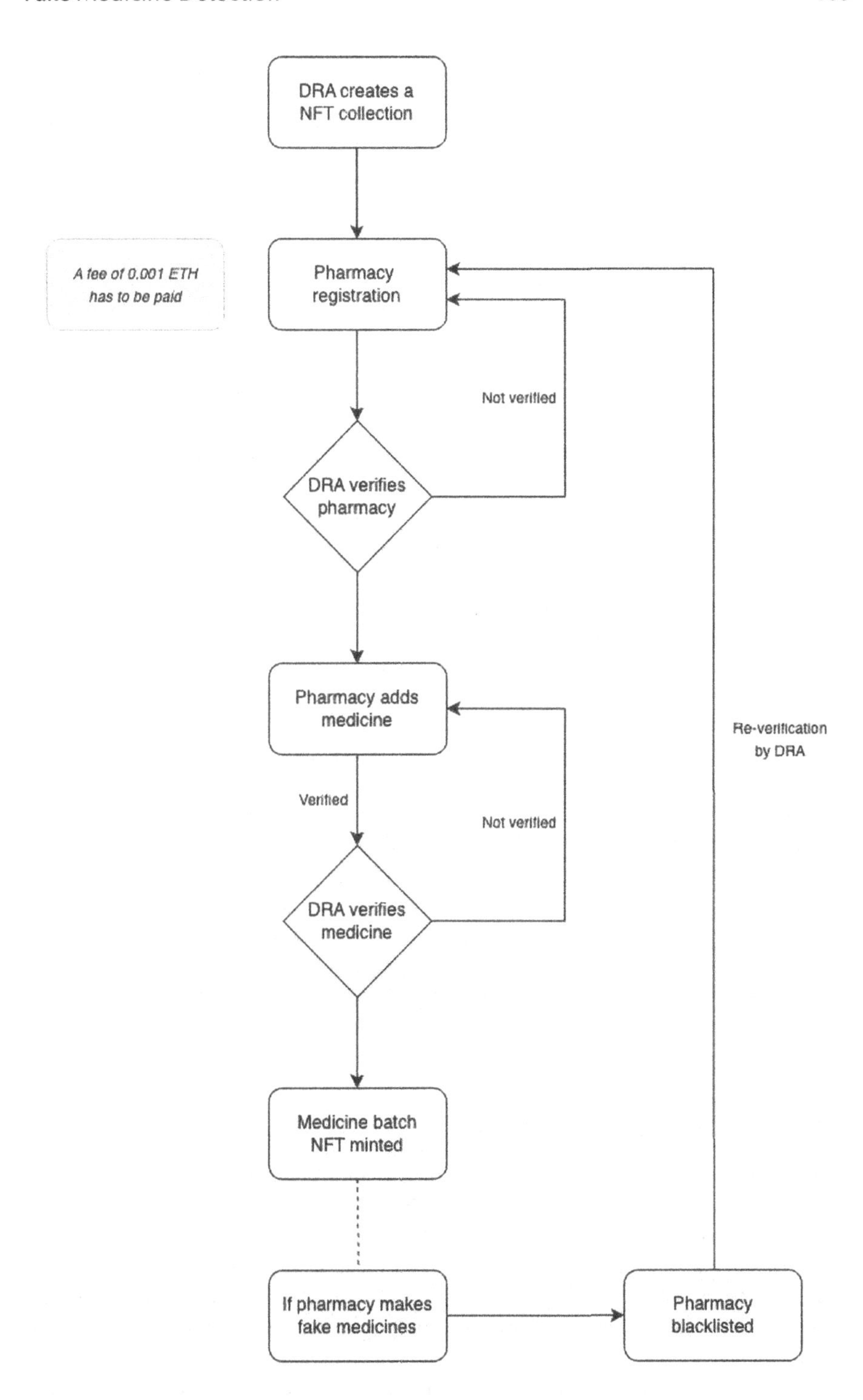

FIGURE 9.3 Flow diagram.

themselves undergo verification by the DRA. A registration fee is required for the pharmacy's enrolment, part of which is allocated to the DRA. When a pharmacy intends to introduce a new medicine, they mint an NFT dedicated to that medicine and associate it with a designated batch number. The DRA subsequently carries out its verification procedures on the newly introduced medicine. In the event of any potentially fraudulent activity concerning either a medicine or a pharmacy, the DRA retains the authority to place the concerned pharmacy on a blacklist.

9.4 CASE STUDY

The case study exemplifies a hypothetical application scenario designed to comprehensively illustrate the potential of the concept. Within the pharmaceutical sector, Company ABC boasts an expansive assortment of medications catering to diverse consumer and retail segments. Faced with the imperative of maintaining competitiveness in a fierce market landscape, Pharmacy ABC must continuously streamline and enhance its operations. The prevalence of cost-effective drugs through conventional distributors presents a challenge, compelling ABC to innovate its processes. This includes developing novel drugs, seeking regulatory approval from the DRA, and accommodating customized designs across various production scales.

Although Company ABC thrives, its reach extends beyond local markets to international destinations. Yet, the concerning frequency of tampering and counterfeit incidents poses a significant threat. Despite earnest efforts to optimize their processes, the manual framework in place makes it nearly impossible to ensure end-to-end authenticity monitoring. Moreover, the company faces the potential hazard of being blacklisted should any unethical practices be detected. In this context, the proposed framework emerges as a valuable solution, offering a robust avenue for enhancing connections between Pharmacy ABC, its local and global pharmacy partners, and the vigilant oversight of the DRA.

By embracing this framework, Company ABC envisions bolstering its operational relationships while safeguarding the authenticity of its medicines. The approach also serves as a cost-efficient pathway for the industry to securely establish its presence and list its medicinal products on the marketplace. This comprehensive strategy not only aids in strengthening ties but also aligns with the global mission of combating counterfeit drugs and ensuring patient safety.

9.5 DATA PRIVACY AND SECURITY

The identification of fake medications holds enormous promise for improving medicinal supply chain integrity in the fast-growing environment of BCT and its use in the Metaverse realm. However, in addition to its potential benefits, this novel technique raises major data privacy concerns that must be carefully considered. "Fake Medicine Detection Using Blockchain in Metaverse Domain," our research study, digs into these important challenges and proposes strategic methods to overcome them.

The virtual and augmented reality (AR) settings of the Metaverse bring new layers of data privacy problems. Because our suggested technique relies on blockchain, specifically the polygon Matic blockchain, to confirm the legitimacy of pharmaceuticals, the incorporation of sensitive medical and transactional data raises concerns about confidentiality, integrity, and control. To detect bogus drugs, we created a thorough system. Some of the unique data privacy concerns we face, as well as our solutions, are listed below:

1. **Patent privacy:** Patient data, including medical history and medication information, is maintained on the blockchain in a Metaverse-driven system. It is critical to ensure that this data stays confidential and available only to authorized persons. Our study describes robust proposals for encryption and access control mechanisms for protecting patient privacy inside the blockchain ecosystem.
2. **Sybil attack:** The Sybil attack is an important threat in blockchain networks, where a malicious node may abuse the system by establishing a slew of pseudonymous identities. To address this risk, a robust solution in the form of the RBAC mechanism has been methodically implemented into our architecture. This novel technique uses the polygon addresses of participating organizations to ensure that transactions are only initiated by authorized nodes with certain rights.
3. **Data ownership and control:** Because blockchain is distributed, it poses concerns regarding data ownership and control. Our research study proposes a governance architecture that defines stakeholders' roles and duties, ensuring that data control stays decentralized while conforming to privacy standards.
4. **Smart contract vulnerabilities**: SCs, which are essential to our system, may have weaknesses that undermine data privacy. We used best coding practices to avoid all potential attacks and used openzeppelin SCs, emphasizing strict security practices to reduce any risks.
5. **Decisive power to users:** The system empowers clients to cast votes regarding potential fraudulent activities perpetrated by pharmacies, subject to review by the DRA. The inherent transparency and immutability of the blockchain ensure that pharmacies cannot tamper with their actions once recorded. This approach guarantees accountability and prevents the removal of incriminating evidence. In the event of guilt, the implicated pharmacy will face the consequence of being publicly renounced. This mechanism establishes a secure and unalterable record of accountability, promoting integrity and trust within the pharmaceutical ecosystem.

9.6 RESULTS AND DISCUSSION

Table 9.2 provides a comprehensive representation of the outcomes of our proposed methodology. It demonstrates that our approach not only entails minimal fees but also boasts an exceptional transaction speed. This heightened transaction speed positions our methodology as a formidable competitor in the realm of modern-day transaction systems. The credit for this remarkable efficiency can be attributed to the strategic

TABLE 9.2
Transaction Details

Actions	Tx fees	Tx hash	Validation time
Contract deployment	0.0048 USD	0x687f223b59f31deb3ede6c00e736887aef1a7f820c 1dc9739ccda183d7469b8d	2 s
Add pharmacy	0.000103 USD	0x029378cf29c4e210c611d0f91ad6ba3cb1ca6223d 533dde2a387bdffd41688f7	4 s
Verify pharmacy	0.000066 USD	0x1951be772a003ed8fcd50e485ad9422fd374696538d 085ea504c2b81cd405a04	4 s
Add medicine	0.000065 USD	0x8b13eb8a21e2f323f660c5cae951a04a4a31fd6a5a 026ed58f50394d3644baf2	2 s
Verify medicine	0.000085 USD	0xe658e0d18b9b8715684a2f46b326b488b4d2957a 3faa8a47ed79ad73348eb3f1	2 s
Renounce pharmacy	0.000074 USD	0x2f345ac0ba3044e94aa67e5589c5d96d4ec 163656138e67388a7acae92f997f1	4 s
Withdraw	0.000035 USD	0x9034f12d871ee4a16b8bd7df283a722fc44d0ae20e 06a3a7b165696f457739b7	4 s

integration of the polygon Matic blockchain. The entire system architecture has been meticulously designed with a keen focus on addressing various exceptional cases and potential errors that may arise during its operation. Furthermore, an additional layer of value is introduced through the emission of events, which occur during specific function calls. This event emission enhances the system's informativeness by providing valuable insights and notifications about critical operations.

We have made significant strides in optimizing transaction throughput, substantially reducing associated transaction fees in the process. This achievement can be attributed to the efficiency of the SC we have developed for our methodology. To substantiate our findings, we conducted a comprehensive analysis across various blockchain platforms, comparing the transaction speeds attained by invoking functions within our SC. Notably, our investigations indicate that the polygon blockchain consistently outperforms other blockchains in terms of average transaction speed.

The efficiency of our SC implementation plays a pivotal role in achieving these results. By carefully designing and optimizing the contract, we have minimized transaction processing times, resulting in a smoother and faster transaction experience. This efficiency is particularly crucial in the context of BCT, where transaction speed and cost-effectiveness are essential factors.

To visually represent our findings, Figure 9.4 illustrates the comparative transaction speeds across different blockchains. It underscores the clear advantage of the polygon blockchain in terms of transaction speed, further reinforcing the significance of our research outcomes. It is important to note that our research is underpinned by rigorous methodology and thorough analysis, ensuring the reliability and validity of our conclusions.

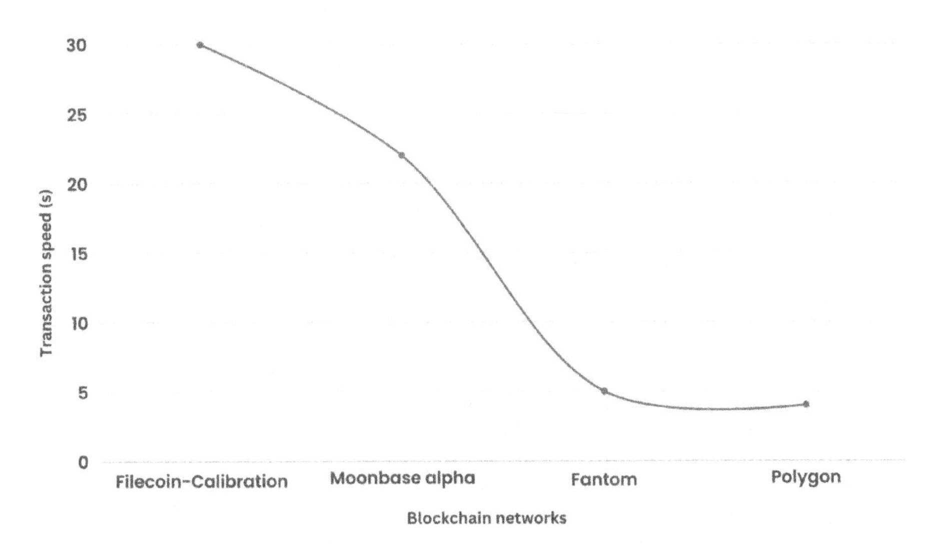

FIGURE 9.4 Transaction throughput of different blockchains.

9.7 CONCLUSIONS AND FUTURE WORK

In this chapter, we offered a complete analysis of the identification of counterfeit pharmaceuticals using BCT integrated innovatively inside the Metaverse domain. Our suggested framework provides a solid response to a significant challenge in the pharmaceutical sector by leveraging the capabilities of the polygon Matic network which drastically reduces the gas fees [29] and QR codes for authenticity verification. We have built a transparent and tamper-proof environment for tracking and validating the validity of medications through a decentralized network including producers, chemists, the DRA, and clients.

Our research underscores the potential of BCT to revolutionize the pharmaceutical supply chain by enhancing traceability, transparency, and accountability. By creating a secure environment where each medicine batch is associated with a unique QR code and verified on the blockchain, we ensure that clients can confidently verify the authenticity of their purchases. This system serves as a deterrent against counterfeit medicines, protecting both consumers and the pharmaceutical industry as a whole.

While our research has yielded promising results, there are several avenues for future exploration and enhancement. First and foremost, further studies could delve into the scalability and performance aspects of the proposed framework, especially as the volume of transactions and participants increases. Additionally, as the Metaverse and blockchain technologies continue to evolve, investigating the integration of advanced features such as AR [30] for enhanced QR code scanning experiences could enrich the user interface. Moreover, the broader implications of our work can extend to addressing other supply chain challenges beyond counterfeit medicines.

REFERENCES

1. A. K. Bapatla, et al., "PharmaChain: A blockchain to ensure counterfeit-free pharmaceutical supply chain," IET Network, vol. 12, no. 2, pp. 53–76, 2023. https://doi.org/10.1049/ntw2.12041
2. V. Rees "The Impact of Counterfeit Drugs in South and South-east Asia". https://www.europeanpharmaceuticalreview.com/article/92194/the-impact-of-counterfeit-drugs-in-south-and-south-east-asia
3. S. Ali, Abdullah, T. P. T. Armand, A. Athar, A. Hussain, M. Ali, M. Yaseen, M.-I. Joo and H.-C. Kim, "Metaverse in healthcare integrated with explainable AI and blockchain: Enabling immersiveness, ensuring trust, and providing patient data security," Sensors, vol. 23, p. 565, 2023. https://doi.org/10.3390/s23020565
4. T. R. Gadekallu, T. Huynh-The, W. Wang, G. Yenduri, P. Ranaweera, Q. V. Pham, D. B. da Costa and M. Liyanage, "Blockchain for the Metaverse: A review," Future Generation Computer Systems (2023).
5. H. Guo and X. Yu, "A survey on blockchain technology and its security," Blockchain: Research and Applications, vol. 3, no. 2, 2022, p. 15.100067, ISSN 2096-7209. https://doi.org/10.1016/j.bcra.2022.100067.
6. A. Haleem, M. Javaid, R. P. Singh, R. Suman and S. Rab, "Blockchain technology applications in healthcare: An overview," International Journal of Intelligent Networks, vol. 2, pp. 130–139, 2021.
7. K. S. Ziavrou, S. Noguera and V. A. Boumba, "Trends in counterfeit drugs and pharmaceuticals before and during COVID-19 pandemic," Forensic Science International, vol. 338, 2022, p. 111382, ISSN 0379-0738. https://doi.org/10.1016/j.forsciint.2022.111382
8. A. Shamsuzzoha, E. Ndzibah and K. Kettunen, "Data-driven sustainable supply chain through centralized logistics network: Case study in a Finnish pharmaceutical distributor company," Current Research in Environmental Sustainability, vol. 2, p. 100013, 2020. https://doi.org/10.1016/j.crsust.2020.100013
9. M. Humayun, N. Z. Jhanjhi, M. Niazi, F. Amsaad and I. Masood, "Securing drug distribution systems from tampering using blockchain," Electronics, vol. 11, p. 1195, 2022. https://doi.org/10.3390/electronics11081195
10. M. Attaran, "Blockchain technology in healthcare: Challenges & opportunities," International Journal of Healthcare Management, vol. 15, no. 1, pp. 70–83, 2022. doi: 10.1080/20479700.2020.1843887.
11. H. Xu, Z. Li, Z. Li, X. Zhang, Y. Sun and L. Zhang, "Metaverse native communication: A blockchain and spectrum prospective," in 2022 IEEE International Conference on Communications Workshops (ICC Workshops), Seoul, Korea, Republic of, 2022, pp. 7–12. doi: 10.1109/ICCWorkshops53468.2022.9814538.
12. B. Lashkari and P. Musilek, "A comprehensive review of blockchain consensus mechanisms," in IEEE Access, vol. 9, pp. 43620–43652, 2021. doi: 10.1109/ACCESS.2021.3065880.
13. A. Back, Hashcash—A Denial of Service Counter-measure, 2002. http://www.hashcash.org/hashcash.pdf
14. Ethereum, Introduction to Smart Contracts, ethereum.org. https://ethereum.org/en/developers/docs/smart-contracts
15. A. Pasdar, Y. C. Lee and Z. Dong, "Connect API with blockchain: A survey on blockchain oracle implementation," ACM Computing Surveys, vol. 55, no. 10, pp. 1–39, 2023. doi: 10.1145/3567582.
16. M. Mettler, "Blockchain technology in healthcare: The revolution starts here," in 2016 IEEE 18th International Conference on e-Health Networking, Applications and Services (Healthcom), Munich, Germany, 2016, pp. 1–3. doi: 10.1109/HealthCom.2016.7749510.

17. K. K. R. Choo, T. McGhin, C. Z. Liu and D. He, "Blockchain in healthcare applications: Research challenges and opportunities," Journal of Network and Computer Applications, vol. 135, pp. 62–75, 2019. doi: 10.1016/j.jnca.2019.02.027.
18. A. Z. Junejo, A. A. Siyal, M. Zawish, K. Ahmed, A. Khalil and G. Soursou, "Applications of blockchain technology in medicine and healthcare: Challenges and future perspectives," Cryptography, vol. 3, no. 1, p. 3, 2019. doi: 10.3390/cryptography3010003.
19. T. Kumar, V. Ramani, I. Ahmad, A. Bracken, E. Harjula and M. Ylianttila, "Blockchain utilization in healthcare: Key requirements and challenges," in IEEE 20th International Conference on e-Health Networking, Applications and Services, Czech Republic, 2018, pp. 1–7.
20. A. Haleem, M. Javaid, R. P. Singh, R. Suman and S. Rab, "Blockchain technology applications in healthcare: An overview," International Journal of Intelligent Networks, vol. 2, pp. 130–139, 2021. doi: 10.1016/j.ijin.2021.09.005.
21. A. Hasselgren, K. Kralevska, D. Gligoroski, S. A. Pedersen and A. Faxvaag, "Blockchain in healthcare and health sciences—A scoping review," International Journal of Medical Informatics, vol. 134, p. 104040, 2020. doi: 10.1016/j.ijmedinf.2019.104040.
22. X. Liang, J. Zhao, S. Shetty, J. Liu and D. Li, "Integrating blockchain for data sharing and collaboration in mobile healthcare applications," in IEEE 28th Annual International Symposium on Personal, Indoor, and Mobile Radio Communications, Canada, 2017, pp. 1–5.
23. A. Dwivedi, G. Srivastava, S. Dhar and R. Singh, "A decentralized privacy-preserving healthcare blockchain for IoT," Sensors, vol. 19, no. 2, p. 326, 2019. https://dx.doi.org/10.3390/s19020326
24. K. N. Griggs, O. Ossipova, C. P. Kohlios, A. N. Baccarini, E. A. Howson and T. Hayajneh, "Healthcare blockchain system using smart contracts for secure automated remote patient monitoring," Journal of Medical Systems, vol. 42, no. 7, PP:1-7, 2018. doi: 10.1007/s10916-018-0982-x.
25. I. Haq and M. Olivier, "Blockchain technology in pharmaceutical industry to prevent counterfeit drugs," International Journal of Computer Applications, vol. 180, pp. 8–12, 2018.
26. I. Radanović and R. Likic, "Opportunities for use of blockchain technology in medicine," Applied Health Economics and Health Policy, vol. 16, no. 5, pp. 583–590, 2018. doi: 10.1007/s40258-018-0412-8.
27. N. Saxena, I. Thomas, P. Gope, P. Burnap and N. Kumar, "PharmaCrypt: Blockchain for critical pharmaceutical industry to counterfeit drugs," IEEE Computer, vol. 53, no. 7, pp. 29–44, 2020. doi: 10.1109/mc.2020.2989238.
28. S. Khan, F. Loukil, C. Ghedira-Guegan, E. Benkhelifa and A. Bani-Hani, "Blockchain smart contracts: Applications, challenges, and future trends," Peer-to-peer Networking and Applications, vol. 14, no. 5, 2901–2925, 2021. doi: 10.1007/s12083-021-01127-0.
29. Cointelegraph. Cointelegraph Bitcoin & Ethereum Blockchain News. Cointelegraph. https://cointelegraph.com/learn/polygon-blockchain-explained-a-beginners-guide-to-matic
30. S. Rostami, and M. Martin. "The Metaverse and beyond: Implementing advanced multiverse realms with smart wearables". IEEE Access 10 (2022), pp: 110796–110806.
31. A. Bandyopadhyay, A. Sarkar, S. Swain, D. Banik, A. E. Hassanien, S. Mallik, A. Li and H. Qin, "A game-theoretic approach for rendering immersive experiences in the Metaverse," Mathematics, vol. 11, no. 6, p. 1286, 2023.
32. A. Sihna, H. Raj, R. Das, A. Bandyopadhyay, S. Swain and S. Chakrborty, "Medical education system based on Metaverse platform: A game theoretic approach," in IEEE 4th International Conference on Intelligent Engineering and Management (ICIEM 2023), 2023, pp. 1–6.
33. P. Gupta, K. Bhadani, A. Bandyopadhyay, D. Banik and S. Swain, "Impact of Metaverse in the near 'future'," in IEEE 4th International Conference on Intelligent Engineering and Management (ICIEM 2023), 2023, pp. 1–6.

10 Smart Healthcare Services Employing Quantum Internet of Things on Metaverse

Kartick Sutradhar
Department of Computer Science and Engineering,
IIIT Sri City, Chittor, India

Ranjitha Venkatesh
Department of Computer Science and Engineering,
GITAM School of Technology Bangalore, India

Priyanka Venkatesh
Department of Computer Science and Engineering,
Presidency University, Bangalore, India

10.1 INTRODUCTION

The Quantum Internet of Things (QIoT) represents a transformative paradigm that combines the principles of quantum mechanics with the interconnectedness of the Internet of Things (IoT) [1]. This emerging technological frontier aims to revolutionize the way we gather, process, and communicate information by harnessing the unique properties of quantum systems, such as superposition and entanglement. QIoT envisions a network where quantum-enabled devices and sensors can securely exchange and process data with unprecedented levels of privacy, efficiency, and accuracy. By leveraging quantum entanglement for secure communication and quantum sensors for ultra-sensitive measurements, QIoT has the potential to enable breakthrough applications in fields ranging from quantum-enhanced sensing and cryptography to distributed quantum computing. However, QIoT also presents substantial technical challenges, including maintaining quantum coherence over long distances, developing reliable quantum communication protocols, and integrating quantum hardware with traditional IoT infrastructure. As research and development in this field progress, QIoT holds the promise of transforming industries and paving the way for a new era of interconnected, quantum-powered devices.

The integration of QIoT into the realm of smart healthcare holds remarkable significance, poised to redefine the landscape of medical technology and patient care. By harnessing the principles of quantum mechanics, QIoT offers a paradigm shift

 DOI: 10.1201/9781003449256-10

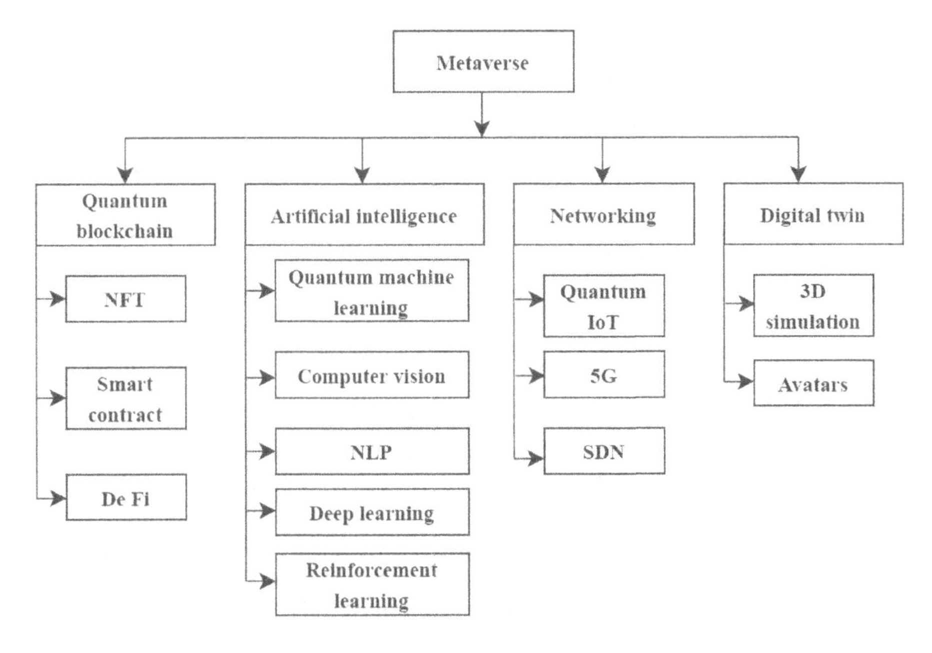

FIGURE 10.1 Different technologies of the Metaverse.

in healthcare innovation. Quantum-enabled sensors and communication protocols empower unprecedented precision in monitoring patients' vital signs, enabling early detection of anomalies and personalized treatment strategies in Metaverse [2–5]. The inherent security of quantum cryptography ensures the utmost confidentiality of sensitive patient data, fostering trust in telemedicine and remote health monitoring. Furthermore, quantum computing expedites intricate simulations for drug discovery and genomic analysis, accelerating the development of tailored therapies based on individual genetic profiles. QIoT's potential to enhance medical imaging accuracy, refine disease modeling, and safeguard data integrity in health records holds the promise of revolutionizing healthcare, ushering in an era of unparalleled insights, resilience, and efficiency. Figure 10.1 shows different technologies of the Metaverse.

10.2 EXISTING ENABLING TECHNOLOGIES FOR SMART HEALTHCARE

10.2.1 Sensors

In the context of smart healthcare, sensors play a pivotal role in transforming traditional medical practices into more efficient, data-driven, and patient-centric approaches. These miniature devices are capable of capturing a wide array of physiological, environmental, and behavioral data, enabling real-time monitoring and analysis of patient conditions. Sensors are seamlessly integrated into wearable devices, medical equipment, and even the built environment, creating a network of interconnected data sources that empower healthcare professionals to make informed decisions and provide personalized care. Wearable sensors, for instance, offer continuous monitoring of vital signs such as heart rate, blood pressure, and body temperature, enabling early detection of anomalies and allowing for timely interventions. Moreover, they facilitate remote patient monitoring, extending medical supervision beyond the confines of healthcare facilities and enhancing patient independence [6].

Incorporating sensors into medical equipment and implants enables precise diagnostics and treatments. Imaging sensors provide detailed insights into internal structures and aid in minimally invasive procedures, reducing patient discomfort and recovery times. Similarly, smart implants equipped with sensors can monitor postoperative progress and adjust treatment parameters in real-time, optimizing outcomes and patient comfort. Beyond individual patient care, sensors contribute to the creation of smart healthcare environments. Environmental sensors monitor air quality, temperature, and humidity, ensuring optimal conditions for patients and healthcare workers. Additionally, location-based sensors enhance hospital efficiency by tracking the movement of equipment, personnel, and patients, streamlining workflows and resource allocation. Despite these advancements, challenges such as data security, interoperability, and patient privacy remain crucial considerations in the implementation of sensor-based solutions. As smart healthcare continues to evolve, sensors stand as essential components that drive the integration of technology and medicine, ultimately leading to more proactive, personalized, and effective patient care.

10.2.2 Big Data

In the realm of smart healthcare, the concept of big data has emerged as a transformative force, revolutionizing the way healthcare systems operate and patients receive care. Big data refers to the massive volumes of diverse and complex data generated from various sources, including medical records, wearable devices, diagnostic imaging, and even social media. Leveraging advanced analytics and machine learning techniques, healthcare professionals can extract valuable insights from this vast pool of information, leading to more accurate diagnoses, personalized treatment plans, and improved patient outcomes. Big data analytics enables the identification of patterns, trends, and correlations that might otherwise go unnoticed. By analyzing large datasets, healthcare providers can enhance clinical decision-making, predict disease outbreaks, and optimize resource allocation within hospitals and clinics. Furthermore, the integration of big data with electronic health records (EHRs) facilitates comprehensive patient profiles, enabling physicians to make well-informed decisions based on a holistic understanding of a patient's medical history and conditions [7].

In the context of research and drug development, big data plays a pivotal role in accelerating scientific advancements. Researchers can analyze genetic data, molecular interactions, and clinical trial outcomes on a massive scale, leading to the discovery of novel treatments and therapeutic targets. This data-driven approach expedites the drug discovery process and holds the potential to revolutionize precision medicine by tailoring treatments to individual patients based on their genetic makeup and medical history. However, the utilization of big data in healthcare also raises significant challenges, including data privacy, security, and ethical considerations. Safeguarding patient information and ensuring compliance with regulatory standards are essential to maintain trust in the healthcare system [8].

10.2.3 Artificial Intelligence

Artificial intelligence (AI) stands as a cornerstone of innovation in smart healthcare, revolutionizing the industry by augmenting human capabilities and driving unprecedented levels of efficiency, accuracy, and patient-centered care. AI encompasses a range of technologies, including machine learning, natural language processing, and computer vision, which collectively empower healthcare systems to process, analyze, and interpret vast amounts of data with remarkable speed and precision [9].

One of the most impactful applications of AI in smart healthcare is medical diagnostics. AI-powered algorithms can rapidly analyze medical images, such as X-rays, MRIs, and CT scans, to detect anomalies and assist radiologists in making more accurate diagnoses. This not only expedites the diagnostic process but also reduces the likelihood of human errors, ultimately leading to improved patient outcomes. Furthermore, AI-driven predictive analytics contribute to proactive and personalized patient care. By analyzing historical patient data, AI algorithms can forecast disease progression and recommend tailored treatment plans. This proactive approach enables healthcare providers to intervene at earlier stages, potentially preventing adverse health events and optimizing the allocation of resources.

In clinical settings, AI-powered virtual health assistants offer real-time support to both patients and healthcare professionals. These AI-driven platforms can answer patient queries, provide medication reminders, and even assist doctors in diagnosing and prescribing treatments based on symptoms and medical history. While AI holds immense promise for smart healthcare, it also presents challenges related to data privacy, bias, and ethical considerations. The responsible development and deployment of AI systems require robust safeguards to ensure patient confidentiality and fairness. AI's integration into smart healthcare marks a paradigm shift in how medical services are delivered and experienced. By leveraging AI's analytical prowess, healthcare systems can elevate precision, efficiency, and patient satisfaction, leading to a future where technology and compassionate care converge to transform the healthcare landscape [10].

10.2.4 Wireless Communication Networks

Wireless communication networks have emerged as an integral component of smart healthcare, revolutionizing the way medical information is transmitted, shared, and utilized. These networks enable seamless and real-time connectivity among medical devices, wearable sensors, EHRs, and healthcare professionals, fostering a dynamic ecosystem that enhances patient care and operational efficiency. The advent of 5G technology has significantly accelerated the capabilities of wireless communication networks in healthcare. With ultralow latency and high data transfer speeds, 5G enables the instantaneous transmission of critical medical data, such as high-resolution imaging and real-time patient monitoring, even in remote or resource-limited settings. This newfound connectivity empowers healthcare practitioners to make informed decisions rapidly and collaborate seamlessly across different locations. Moreover, wireless networks play a pivotal role in remote patient monitoring and telemedicine. Patients can use wearable devices equipped with sensors to track vital signs, activity levels, and medication adherence. These devices transmit data wirelessly to healthcare providers, allowing for continuous monitoring and timely interventions. Telemedicine, enabled by robust wireless networks, enables virtual consultations, remote diagnoses, and prescription management, extending medical services to individuals who may have limited access to healthcare facilities [11].

The integration of wireless communication networks also enhances the efficiency of healthcare operations. Hospital staff can use wireless devices to access patient information, update medical records, and coordinate care in real-time, streamlining workflows and reducing administrative burdens. However, the implementation of wireless networks in smart healthcare also presents challenges. Data security, privacy concerns, and network reliability are paramount considerations to ensure the confidentiality of patient information and the uninterrupted operation of critical medical systems. The wireless communication networks serve as the backbone of smart healthcare, enabling the interconnectedness of devices, data, and healthcare professionals. As technology continues to evolve, these networks will play an increasingly crucial role in shaping the future of healthcare delivery, fostering improved patient outcomes and enhancing the overall quality of medical services [12].

10.2.5 Cloud and Edge Computing

Cloud and edge computing have emerged as foundational pillars in the realm of smart healthcare, revolutionizing the way medical data is processed, stored, and accessed. These computing paradigms offer distinct advantages that collectively enhance the efficiency, scalability, and accessibility of healthcare services. Cloud computing provides a robust and centralized platform for storing and processing vast amounts of medical data, ranging from EHRs and diagnostic images to patient histories and treatment plans. Healthcare institutions can leverage cloud-based systems to securely store and share information, enabling seamless collaboration among healthcare providers and facilitating access to patient data from anywhere with an internet connection. The scalability of cloud resources ensures that healthcare systems can accommodate growing data volumes and computational demands, while also enabling advanced analytics and machine learning algorithms to derive valuable insights from the collected data. On the other hand, edge computing complements cloud computing by bringing computational power closer to the data source, minimizing latency and enabling real-time processing. In smart healthcare, edge devices such as wearable sensors and medical implants can process and analyze data locally, allowing for rapid responses and immediate feedback. This is particularly critical for applications like remote patient monitoring, where timely alerts and interventions can significantly impact patient outcomes. Edge computing also reduces the need for continuous high-bandwidth data transmission to the cloud, alleviating network congestion and enhancing overall system efficiency [13].

The integration of cloud and edge computing in smart healthcare presents opportunities for hybrid architectures, where critical data processing occurs at the edge while leveraging cloud resources for complex analytics and long-term storage. However, challenges related to data privacy, security, and interoperability require careful consideration. Ensuring that patient data remains protected throughout its journey between edge devices and cloud servers is of paramount importance [14]. The cloud and edge computing form a dynamic and symbiotic duo in the landscape of smart healthcare. By combining the strengths of centralized cloud resources with the immediacy of edge processing, healthcare systems can achieve enhanced data-driven insights, improved patient care, and streamlined operational processes, ultimately contributing to a more efficient and patient-centric healthcare ecosystem.

10.2.6 Immersive Technology

Immersive technology has emerged as a transformative force in the realm of smart healthcare, redefining the ways in which medical professionals deliver care, patients engage with their health, and medical education and training are conducted. Immersive technologies, including virtual reality (VR) and augmented reality (AR), create immersive and interactive environments that enable users to experience medical scenarios and information in unprecedented ways. In smart healthcare, immersive technology finds application across various domains. Medical education and training benefit immensely from immersive simulations, allowing students and practitioners to engage in realistic medical scenarios without risk to patients. Surgeons can practice complex procedures in a virtual environment, refining their skills and enhancing surgical precision before entering the operating room. Likewise, medical students can explore three-dimensional anatomical models, enhancing their understanding of human physiology and pathology. Patient care is also transformed through immersive technology. VR can offer pain relief and relaxation during medical procedures, mitigating patient anxiety and discomfort. AR overlays relevant patient information onto a surgeon's field of view during procedures, enabling them to access critical data without diverting their attention from the patient. Moreover, immersive experiences can aid in patient education, helping individuals better comprehend their diagnoses and treatment options [15].

In research and diagnosis, immersive technology contributes to data visualization and analysis. Complex medical data, such as molecular structures or diagnostic images, can be represented in three-dimensional space, allowing researchers to gain deeper insights and make more informed decisions. Despite the immense potential, challenges such as content quality, standardization, and accessibility need to be addressed to fully integrate immersive technology into smart healthcare. Additionally, ensuring patient privacy and data security remains paramount. The immersive technology is poised to reshape the landscape of smart healthcare, enhancing medical education, patient care, and research. By merging the virtual and physical worlds, immersive technology opens up new avenues for innovation and collaboration, fostering a future where healthcare is not only advanced but also more engaging and personalized [16].

10.3 ENABLING TECHNOLOGIES OF THE METAVERSE FOR HEALTHCARE

10.3.1 Extended Reality

Extended reality (XR) stands at the forefront of the Metaverse's potential for revolutionizing healthcare, offering a seamless integration of VR, AR, and mixed reality technologies. As the boundaries between the physical and digital worlds blur, XR creates immersive and interactive experiences that hold immense promise for healthcare applications. In the Metaverse for healthcare, XR transforms medical practices across diverse domains. Patient engagement and education are elevated to new heights as XR allows individuals to visualize complex medical concepts in three-dimensional space [17]. Patients can explore virtual models of their own anatomy, gaining a deeper understanding of their conditions and treatment plans. Additionally, XR-enhanced telemedicine enables remote consultations where healthcare providers can virtually examine patients, enhancing the quality and effectiveness of long-distance medical care. Training and medical education are revolutionized through XR by enabling healthcare professionals to practice procedures and surgeries in virtual environments. Medical students can gain hands-on experience in simulated scenarios, fostering skill development and confidence before encountering real patients. XR also facilitates interdisciplinary collaboration, allowing experts from various fields to work together within a shared virtual space, transcending geographical limitations.

Diagnosis and treatment benefit from XR's capabilities as well. Radiologists can manipulate and analyze medical images in three dimensions, potentially uncovering subtle details that might be missed in traditional 2D interpretations. Surgeons can overlay vital patient information onto their field of view through AR, enhancing surgical precision and reducing the need for distraction [18]. However, while XR's potential is vast, challenges such as hardware accessibility, content development, and user acceptance need careful consideration. Ensuring that XR experiences are inclusive and user-friendly, while maintaining patient privacy and data security, is crucial for realizing its full potential in healthcare. The integration of XR within the Metaverse holds transformative potential for healthcare by offering immersive, interactive, and collaborative experiences. As XR continues to evolve, it has the capacity to redefine medical education, patient care, and diagnostics, ushering in a new era of innovation and enhanced outcomes in the healthcare landscape.

10.3.2 Blockchain

Blockchain technology stands as a foundational pillar within the Metaverse, holding the potential to revolutionize healthcare by addressing critical challenges related to data security, interoperability, and patient privacy. In the context of the Metaverse for healthcare, blockchain offers a decentralized and tamper-proof ledger that securely records and shares medical information across disparate systems and stakeholders. This technology ensures that patient data remain immutable and transparent, enhancing trust among patients, healthcare providers, and researchers. By utilizing blockchain, healthcare institutions can create a unified and secure ecosystem where patient records, medical histories, and treatment plans are seamlessly shared and updated. This interoperable environment streamlines the exchange of information, reducing administrative burdens and minimizing the risk of errors that can arise from fragmented data systems. Additionally, blockchain enables patients to have greater control over their health data, granting them the ability to grant access and revoke permissions, ultimately empowering individuals to actively participate in their care decisions [19].

Blockchain's potential for secure data exchange extends to medical research and clinical trials within the Metaverse. Researchers can securely access and share anonymized patient data, accelerating the discovery of new treatments and therapies. Smart contracts on the blockchain can automate and ensure compliance with research protocols, enhancing transparency and reducing administrative overhead. Despite its promise, integrating blockchain into the Metaverse for healthcare requires addressing challenges such as scalability, regulatory compliance, and standardization. Ensuring that blockchain solutions are scalable and capable of handling the vast amounts of medical data generated within the Metaverse is essential. Moreover, navigating the complex regulatory landscape and establishing industry-wide standards are crucial steps to enable widespread adoption. The blockchain technology holds the key to unlocking a secure, transparent, and interoperable Metaverse for healthcare [20]. By enhancing data security, patient privacy, and collaborative research, blockchain has the potential to reshape healthcare practices and empower individuals within the Metaverse, ultimately leading to improved patient outcomes and transformative advancements in medical research and practice.

10.3.3 Quantum Internet of Things

The integration of QIoT within the Metaverse has the potential to redefine the landscape of healthcare by leveraging the unique capabilities of quantum computing and communication. QIoT extends the concept of traditional IoT by harnessing quantum entanglement and superposition to create a highly interconnected and secure network of devices and sensors. Within the Metaverse for healthcare, QIoT offers unprecedented levels of data processing, encryption, and communication that can significantly enhance patient care, medical research, and diagnostics. In the realm of healthcare, QIoT enables real-time monitoring of patient vitals and environmental factors with unparalleled precision. Quantum sensors can detect subtle changes in biological markers and provide data at quantum-limited levels of accuracy, enabling early detection of diseases and enabling proactive interventions. Furthermore, QIoT enhances telemedicine by ensuring the secure transmission of sensitive patient data, safeguarding privacy, and fostering trust between patients and healthcare providers. QIoT's quantum computing capabilities hold transformative potential for medical research within the Metaverse. Complex simulations and molecular modeling required for drug discovery and personalized medicine can be accelerated exponentially, leading to the rapid development of innovative treatments and therapies. Quantum-enhanced machine learning algorithms can decipher intricate patterns within massive healthcare datasets, leading to more accurate diagnoses and tailored treatment plans [21]. However, the realization of QIoT's potential in the Metaverse for healthcare also faces formidable challenges. Quantum computing technology is still in its early stages of development and scalability, and practical implementation remains a significant hurdle. Moreover, the integration of quantum communication within the existing infrastructure requires careful consideration of compatibility and security. The QIoT presents a paradigm-shifting opportunity within the Metaverse for healthcare, offering quantum-powered capabilities that extend beyond the confines of classical computing and communication. While challenges persist, the potential for QIoT to revolutionize patient care, medical research, and diagnostics within the Metaverse underscores its significance as a cutting-edge technology with transformative implications for the future of healthcare [22].

10.3.4 Digital Twin

The incorporation of Digital Twin technology into the Metaverse holds immense promise for transforming healthcare by creating virtual replicas of both individual patients and entire healthcare systems. A Digital Twin in the Metaverse for healthcare represents a comprehensive and dynamic model that mirrors real-world patient physiology, medical history, and treatment responses. This technology enables healthcare professionals to simulate and analyze various scenarios, optimizing personalized treatment plans and interventions. Digital Twins offer a novel approach to patient-centric care within the Metaverse. By integrating data from wearable sensors, EHRs, and medical imaging, a patient's Digital Twin provides a holistic and real-time representation of their health status. Healthcare providers can monitor physiological changes, predict potential health issues, and devise personalized interventions that are tailored to the unique characteristics of the patient's Digital Twin. This fosters proactive and preventive care strategies, ultimately leading to improved patient outcomes. Moreover, Digital Twins extend their impact beyond individual patients to healthcare infrastructure and systems. Within the Metaverse, healthcare institutions can create Digital Twins of their facilities, enabling real-time monitoring of equipment, resource allocation, and patient flow. These virtual replicas facilitate operational optimization, predictive maintenance, and resource allocation, ensuring that healthcare systems run efficiently and deliver high-quality care [23].

The integration of Digital Twins in the Metaverse for healthcare also poses challenges such as data privacy, integration of diverse data sources, and maintaining accurate and up-to-date models. Ensuring the security and privacy of patient data within the Digital Twin ecosystem is of paramount importance to build and maintain patient trust. Digital Twins represent a transformative concept within the Metaverse for healthcare, offering a dynamic, real-time, and personalized representation of patients and healthcare systems. By leveraging Digital Twin technology, the healthcare industry within the Metaverse stands to revolutionize patient care, enhance operational efficiency, and drive innovation in medical research and practice.

10.3.5 Human–Computer Interaction

Human–computer interaction (HCI) plays a pivotal role in shaping the user experience and accessibility of the Metaverse for healthcare, facilitating seamless interactions between individuals and the digital healthcare ecosystem. In the context of the Metaverse, HCI is a multidimensional interface that enables users, including healthcare professionals, patients, and researchers, to engage with a complex amalgamation of virtual environments, data, and interconnected devices. HCI within the Metaverse for healthcare focuses on designing intuitive interfaces that enhance user engagement and streamline workflows [24]. Healthcare professionals can use gesture-based controls, voice commands, and haptic feedback to navigate virtual patient records, visualize medical data, and conduct simulations. Patients can interact with their Digital Twins or engage in telemedicine consultations using natural and familiar modes of interaction, fostering a sense of empowerment and control over their health.

Personalization is a central theme in HCI within the Metaverse for healthcare. Tailoring interfaces to individual user preferences and needs ensures that interactions are efficient and relevant. HCI also extends to accessibility, aiming to make the Metaverse inclusive for individuals with diverse abilities, thereby ensuring equitable access to healthcare services [25]. Furthermore, HCI within the Metaverse for healthcare is pivotal for facilitating interdisciplinary collaborations. Healthcare professionals, researchers, data scientists, and engineers can interact in shared virtual spaces, enabling real-time discussions, data analysis, and collaborative problem-solving. This interconnectedness catalyzes innovation, enabling swift knowledge exchange and propelling advancements in medical research and clinical practice. However, HCI within the Metaverse for healthcare faces challenges such as designing interfaces that effectively manage information overload and integrating new interaction paradigms seamlessly into existing workflows. Ensuring data privacy, security, and maintaining ethical considerations in the Metaverse's interactive experiences also remain critical aspects of HCI. HCI within the Metaverse for healthcare is a dynamic and evolving field that shapes the way individuals interact with the digital healthcare ecosystem. By prioritizing intuitive design, personalization, and accessibility, HCI enhances the Metaverse's potential to empower healthcare stakeholders, foster collaborations, and ultimately enhance patient care, research, and medical education.

10.4 COMPONENTS OF QUANTUM INTERNET OF THINGS

QIoT is a multidisciplinary concept that combines elements from quantum physics, information technology, and IoT infrastructure. It involves the integration of quantum technologies into traditional IoT systems to create a more secure, efficient, and capable network. The components of QIoT include:

a. Quantum sensors: These specialized devices leverage quantum effects to measure physical quantities with unprecedented accuracy and sensitivity. Quantum sensors can be used to monitor various parameters such as temperature, pressure, magnetic fields, and chemical compositions. In healthcare applications, quantum sensors could play a role in real-time patient monitoring and early disease detection.
b. Quantum communication: Quantum communication protocols utilize the principles of quantum mechanics, such as entanglement and superposition,

to ensure secure and tamper-proof data transmission. Quantum key distribution (QKD) is a fundamental component, enabling encryption keys to be securely shared between devices, making communications virtually unbreakable.

c. Quantum computing: Quantum computers have the potential to solve complex problems much faster than classical computers. In QIoT, quantum computing could be used for tasks such as optimizing logistics, simulating complex biological systems for drug discovery, and solving optimization problems in various domains.
d. Quantum cryptography: Quantum cryptography uses quantum properties to create unbreakable encryption methods. QKD, for instance, can provide secure communication channels for transmitting sensitive data between devices within the IoT network [26].
e. Quantum repeaters: These devices are crucial for maintaining quantum entanglement over long distances in quantum communication networks. They help overcome the challenge of quantum information degradation due to environmental factors.
f. Quantum software and algorithms: QIoT requires specialized algorithms and software tailored to quantum computing hardware. These algorithms optimize the use of quantum resources and perform tasks that would be infeasible for classical computers.
g. IoT devices and infrastructure: Traditional IoT devices, such as sensors, actuators, and communication modules, form the foundation of QIoT. These devices collect and transmit data to the quantum-enhanced components of the network.
h. Data processing and analytics: Quantum-enhanced data processing and analytics can extract insights from large and complex datasets, contributing to real-time decision-making and predictive modeling in various applications, including healthcare, logistics, and environmental monitoring.
i. Security protocols: QIoT employs advanced security protocols to safeguard sensitive data from quantum and classical cyber threats, ensuring the integrity, confidentiality, and authenticity of information exchanged within the network.

QIoT can be applied across various sectors, including smart healthcare, transportation, energy management, environmental monitoring, and more, where its capabilities can lead to substantial improvements in efficiency, security, and data processing. These components collectively form the foundation of QIoT, enabling the creation of a network that merges the power of quantum technologies with the interconnectedness of the IoT, opening up new possibilities for innovation and advancement across diverse industries [27].

10.4.1 Quantum IoT for Secured Healthcare Information Transmission

QIoT presents a transformative solution for ensuring the utmost security in the transmission of healthcare information. By leveraging the principles of quantum mechanics, QIoT offers an unprecedented level of data protection that is fundamentally unbreakable by conventional means. Through quantum communication protocols such as QKD, QIoT establishes encryption keys that are impervious to eavesdropping and hacking attempts, safeguarding sensitive medical records, diagnostic results, and patient information from unauthorized access. Quantum entanglement-based communication ensures that any attempts to intercept or alter data would be instantly detectable, providing an unassailable shield against cyber threats [28]. This level of security not only preserves patient privacy and regulatory compliance but also fosters trust in telemedicine and remote patient monitoring, ultimately advancing healthcare information transmission into an era of inviolable confidentiality and integrity.

10.4.2 Integration with Quantum Internet of Things for Healthcare on Metaverse

The integration of QIoT within the healthcare domain extends its potential impact to the emerging Metaverse, creating a dynamic and secure ecosystem for medical data and services. By intertwining the capabilities of QIoT with the Metaverse's virtual environments, individuals could access personalized healthcare experiences seamlessly. Quantum-secured communication protocols would ensure that sensitive health information shared within the Metaverse remains impervious to unauthorized access or tampering. Real-time monitoring through quantum sensors could provide users with accurate and reliable health data, enhancing virtual health interactions and simulations. Moreover, the computational power of quantum computing could optimize the Metaverse's ability to simulate complex physiological and pathological scenarios, aiding medical research and training [29]. As users navigate this interconnected digital realm, the fusion of QIoT and the Metaverse could revolutionize remote healthcare consultations, medical education, and even contribute to novel avenues of collaborative research, ultimately reshaping the way healthcare is delivered and experienced in this immersive digital frontier.

10.4.3 Quantum Sensing and Imaging for Healthcare Applications on Metaverse

Quantum sensing and imaging, when integrated into healthcare applications within the Metaverse, introduce a new dimension of precision and insight. By leveraging the principles of quantum mechanics, healthcare experiences in the Metaverse can offer unparalleled diagnostic and monitoring capabilities. Quantum sensors, capable of detecting minute changes in physical parameters, enable users to obtain real-time, ultra-sensitive health data in virtual environments. These data, securely transmitted through quantum communication, allow for accurate and continuous remote patient monitoring, bridging the physical and virtual realms seamlessly. Quantum-enhanced imaging techniques, leveraging the phenomenon of entanglement, enable users to visualize intricate biological structures and physiological processes at an unprecedented level of detail. These quantum-enabled advancements facilitate highly accurate medical diagnoses, personalized treatment simulations, and immersive medical training scenarios [30]. Ultimately, the integration of quantum sensing and imaging into healthcare applications within the Metaverse enhances the depth and precision of virtual medical experiences, contributing to a novel frontier of healthcare that transcends traditional boundaries.

10.5 SMART HEALTHCARE APPLICATIONS OF QUANTUM INTERNET OF THINGS ON METAVERSE

The integration of QIoT into smart healthcare applications within the Metaverse offers a transformative avenue for enhancing medical services and experiences. Through the fusion of quantum technologies and virtual environments, the Metaverse becomes a platform for advanced healthcare interactions. Quantum-enabled sensors provide real-time, ultraprecise physiological monitoring, enabling users to engage in immersive self-health assessments and personalized wellness tracking. Secure quantum communication ensures that sensitive medical data shared within the Metaverse remains impervious to breaches, fostering a sense of trust and privacy. Quantum computing's immense computational power facilitates complex simulations and predictive modeling, enabling users to visualize the outcomes of different treatment options and interventions. Additionally, healthcare professionals can leverage the Metaverse for remote consultations, using quantum-enhanced communication to ensure confidential and accurate exchanges of medical information. Through these applications, the convergence of QIoT and the Metaverse in smart healthcare revolutionizes patient engagement, diagnostics, and treatment strategies, offering a novel and empowered approach to virtual healthcare experiences.

10.5.1 Diagnostics and Imaging on Metaverse

The integration of QIoT into diagnostics and imaging within the Metaverse heralds a new era of precision and insight in virtual healthcare experiences. Quantum-enhanced sensors, operating with unprecedented sensitivity, gather real-time physiological data from users, creating a dynamic feedback loop between the virtual environment and the user's actual health metrics. This seamless integration allows for continuous monitoring and early detection of anomalies, contributing to proactive health management. Quantum-enabled imaging techniques further elevate the Metaverse's capabilities, enabling users to visualize complex anatomical structures and physiological processes with unmatched clarity. By harnessing quantum properties like entanglement, these imaging methods offer unparalleled resolution and accuracy, facilitating detailed medical examinations and facilitating immersive educational experiences. Secure quantum communication ensures the privacy and integrity of sensitive medical information shared within the Metaverse, building trust and confidence among users [31]. Through the synergy of QIoT, quantum-enhanced imaging, and the immersive nature of the Metaverse, individuals can engage in comprehensive self-assessments, informed health discussions, and interactive medical simulations, all while redefining the boundaries of virtual healthcare diagnostics and imaging.

10.5.2 Drug Discovery and Development on Metaverse

The integration of QIoT with drug discovery and development within the Metaverse represents a groundbreaking convergence of advanced technologies to accelerate pharmaceutical innovation. Quantum-enabled simulations and computing resources offer a dynamic platform for modeling complex molecular interactions with unprecedented accuracy and efficiency. By harnessing quantum mechanics, researchers can simulate the behavior of molecules, predict drug interactions, and optimize molecular structures in virtual environments, significantly expediting the drug discovery process. The Metaverse's immersive capabilities provide an interactive space where scientists can visualize and manipulate these molecular simulations, fostering collaborative research and creative problem-solving. Secure quantum communication ensures the confidential exchange of proprietary research data and findings, enabling global teams to seamlessly collaborate in real-time. This transformative synergy empowers pharmaceutical researchers to explore an expansive virtual landscape of molecular possibilities, leading to the accelerated development of novel and targeted therapies. Ultimately, the marriage of QIoT and the Metaverse redefines the boundaries of drug discovery, creating an innovative ecosystem that holds the potential to revolutionize pharmaceutical advancements and reshape the future of healthcare [32].

10.5.3 Wearable Health Monitoring Devices on Metaverse

QIoT has ushered in a remarkable era of healthcare innovation, and its integration with wearable health monitoring devices within the Metaverse presents a transformative paradigm for personalized well-being. Quantum-enabled sensors embedded in these wearables offer an unprecedented level of accuracy in capturing physiological data, providing real-time insights into users' health metrics. As these devices seamlessly interface with the Metaverse, users can engage in immersive and interactive health monitoring experiences, visualizing their vital signs and trends in dynamic virtual environments. Secure quantum communication ensures the privacy and integrity of sensitive health data shared between wearables and the Metaverse, fostering a sense of trust and confidence [33]. Moreover, quantum computing's computational power enhances data analytics, enabling wearables to process vast amounts of information and provide personalized health recommendations tailored to individual needs. This integration empowers users to actively participate in their health management, seamlessly transitioning between the physical and virtual realms. Through the convergence of QIoT-enabled wearables and the Metaverse, users can embark on a holistic journey of self-care, leveraging cutting-edge technology to optimize well-being and establish a novel dimension of connected healthcare experiences.

10.5.4 Telemedicine and Remote Healthcare on Metaverse

Telemedicine and remote healthcare are set to undergo a profound transformation through integration with the Metaverse, offering an immersive and interconnected approach to medical consultations and treatment. By merging VR and telemedicine, patients can engage in lifelike interactions with healthcare providers within dynamic virtual environments. The Metaverse's interactive capabilities allow for realistic examination scenarios, enabling physicians to remotely assess patients' conditions and offer accurate diagnoses. Quantum-secured communication ensures the confidentiality of sensitive medical data shared during these virtual consultations, fostering a secure and private environment for healthcare interactions. Patients, regardless of geographical constraints, can access specialized expertise and second opinions seamlessly, leading to more informed medical decisions and comprehensive care. Additionally, the Metaverse's potential for real-time monitoring and visualization empowers patients to actively participate in their treatment plans, enhancing health education and self-management. This convergence of telemedicine and the Metaverse not only expands access to quality healthcare but also introduces a new dimension of patient engagement and empowerment, paving the way for a future where remote healthcare transcends physical limitations and redefines the patient–provider relationship [34].

10.6 ADVANTAGES AND CHALLENGES OF QUANTUM INTERNET OF THINGS IN HEALTHCARE

The integration of QIoT into smart healthcare applications within the Metaverse offers a host of compelling advantages, yet it also presents notable challenges. On the positive side, QIoT brings unparalleled security through quantum cryptography, ensuring the confidentiality and integrity of sensitive medical data exchanged between users and healthcare providers in virtual environments. Quantum-enabled sensors and imaging enhance diagnostics, allowing for real-time, high-precision health monitoring and visualization of intricate anatomical structures [35]. The Metaverse's immersive nature provides a unique platform for interactive patient education, remote consultations, and collaborative medical research. However, challenges persist, including the need for robust and scalable quantum hardware, the development of specialized quantum algorithms for healthcare applications, and the seamless integration of QIoT components with Metaverse platforms. Moreover, ensuring interoperability between various devices and protocols within this complex ecosystem requires careful consideration. Overcoming these challenges is essential to fully realizing the transformative potential of QIoT in smart healthcare on the Metaverse, where the fusion of quantum technologies and VR can reshape the landscape of medical services, diagnostics, and patient engagement.

10.6.1 Security and Privacy Considerations on Metaverse

The integration of QIoT within the Metaverse introduces a new frontier of security and privacy considerations. While QIoT promises unprecedented levels of data protection through quantum cryptography, ensuring the confidentiality and integrity of sensitive information exchanged in virtual healthcare environments, it also raises complex challenges. Quantum-secured communication safeguards against classical cyber threats, yet the Metaverse's intricate interconnectedness demands vigilance against emerging quantum-based attacks that could exploit vulnerabilities. Moreover, the immersive nature of the Metaverse necessitates enhanced measures to protect user identity and maintain data privacy [21]. The integration of personal health data, even within a quantum-secure framework, requires robust consent mechanisms and stringent access controls to prevent unauthorized usage or potential breaches. Striking the delicate balance between leveraging the power of QIoT for advanced healthcare interactions and upholding the highest standards of security and privacy in the Metaverse presents an ongoing endeavor, demanding innovative solutions and a comprehensive approach to address the multifaceted nature of these considerations.

10.6.2 Technological and Implementation Challenges on Metaverse

The convergence of QIoT with the Metaverse introduces a realm of exciting possibilities, but it is not without significant technological and implementation challenges. Foremost among these challenges is the development and deployment of robust quantum hardware that can withstand the demands of real-time, immersive Metaverse experiences. Quantum sensors and communication components must exhibit stability and reliability over extended periods while operating in complex virtual environments [15]. The integration of diverse quantum devices with Metaverse platforms demands seamless interoperability and standardized protocols, a task complicated by the nascent nature of both quantum technologies and Metaverse ecosystems. Additionally, the computational demands of simulating quantum interactions within the Metaverse require scalable quantum computing resources, posing a substantial hurdle given the current limitations in quantum hardware capabilities. The Metaverse's reliance on interconnected networks also exposes vulnerabilities to cyber threats, necessitating novel security architectures that are quantum-resistant. Finally, the user experience within the Metaverse must be intuitive and accessible, requiring careful consideration of user interfaces and interaction paradigms that bridge the gap between complex quantum processes and user-friendly engagement. Addressing these technological and implementation challenges is pivotal to fully harnessing the potential of QIoT on the Metaverse, enabling a seamless fusion of quantum technologies and VR for transformative healthcare experiences.

10.7 FUTURE DIRECTIONS AND EMERGING TRENDS

10.7.1 Roadmap for Quantum Internet of Things in Smart Healthcare

Creating a roadmap for the integration of QIoT in smart healthcare within the Metaverse entails a strategic approach that navigates both technological advancements and implementation milestones. The initial phase involves the refinement and deployment of quantum-enabled sensors and communication protocols, ensuring their compatibility with Metaverse platforms and establishing a secure foundation for data exchange. Concurrently, quantum computing capabilities should be scaled to support complex simulations for drug discovery and personalized treatment simulations. As quantum hardware matures, the next phase focuses on optimizing quantum algorithms specifically tailored to healthcare applications within the Metaverse, enabling real-time monitoring, diagnostics, and collaborative medical research. This roadmap must emphasize the development of quantum-resistant security measures to safeguard against emerging cyber threats, ensuring the privacy and integrity of patient data. Interdisciplinary collaboration between quantum experts, healthcare professionals, and Metaverse developers is pivotal to fine-tuning interfaces, creating immersive healthcare experiences that empower users to actively engage in their well-being. Throughout each phase, ongoing research and development should continually address challenges, including interoperability, standardization, and user adoption [36]. This incremental roadmap strives to synergize the power of QIoT with the immersive potential of the Metaverse, culminating in a transformative landscape where quantum-enhanced healthcare interactions redefine patient care, diagnostics, and medical research on an unprecedented scale.

10.7.2 Potential Impact on the Healthcare Industry

The integration of the healthcare industry with the Metaverse holds immense potential to revolutionize healthcare delivery and patient experiences. As the Metaverse bridges the physical and digital realms, it offers a transformative platform for remote consultations, personalized diagnostics, and interactive health education. Patients can engage in immersive virtual visits, enabling access to specialized medical expertise regardless of geographical constraints. The Metaverse's dynamic simulations and visualizations empower medical professionals to enhance diagnostics, treatment planning, and surgical simulations, leading to more accurate interventions and improved patient outcomes. Collaborative research and medical training are elevated through interactive 3D models and real-time knowledge sharing, accelerating scientific advancements. Moreover, the Metaverse encourages proactive patient engagement, allowing individuals to actively manage their health through real-time monitoring, wellness tracking, and immersive health education experiences [37]. While challenges such as data privacy and technology integration must be addressed, the potential impact of the Metaverse on the healthcare industry is profound, promising a future where healthcare is more accessible, interactive, and tailored to individual needs.

10.8 CONCLUSION

The integration of QIoT within smart healthcare services on the Metaverse marks a remarkable stride toward the future of healthcare. The amalgamation of quantum technologies with VR has unveiled a realm where precision, security, and interactivity converge to redefine patient-centric care. Quantum-enabled sensors provide a granular understanding of health metrics, paving the way for proactive monitoring and personalized interventions. The Metaverse, with its immersive environment, transcends geographical boundaries, enabling seamless telemedicine consultations and collaborative medical research. Quantum cryptography bolsters data security, ensuring patient privacy in virtual interactions. Moreover, quantum computing's computational prowess accelerates drug discovery, treatment optimization, and complex medical simulations. The journey, however, is not devoid of challenges, ranging from quantum hardware reliability to interface design. As this integration evolves, stakeholders must navigate these obstacles to harness the full potential of QIoT within the Metaverse. Ultimately, this visionary synergy stands poised to elevate healthcare into an era where quantum-enhanced experiences on the Metaverse redefine wellness, diagnostics, and patient empowerment, shaping a future that is both technologically advanced and compassionately patient-centered.

REFERENCES

1. A. Hoerbst and E. Ammenwerth, "Electronic health records," Methods of Information in Medicine, vol. 49, no. 04, pp. 320–336, 2010.
2. Y. Wang, Z. Su, N. Zhang, R. Xing, D. Liu, T. H. Luan and X. Shen, "A survey on Metaverse: Fundamentals, security, and privacy," IEEE Communications Surveys & Tutorials, 2022.
3. A. Bandyopadhyay, A. Sarkar, S. Swain, D. Banik, A. E. Hassanien, S. Mallik, A. Li and H. Qin, "A game-theoretic approach for rendering immersive experiences in the Metaverse," Mathematics, vol. 11, no. 6, p. 1286, 2023.
4. A. Sihna, H. Raj, R. Das, A. Bandyopadhyay, S. Swain and S. Chakrborty, "Medical education system based on Metaverse platform: A game theoretic approach," in IEEE 4th International Conference on Intelligent Engineering and Management (ICIEM 2023), 2023, pp. 1–6.
5. P. Gupta, K. Bhadani, A. Bandyopadhyay, D. Banik and S. Swain, "Impact of Metaverse in the near 'future'," in IEEE 4th International Conference on Intelligent Engineering and Management (ICIEM 2023), 2023, pp. 1–6.
6. W. J. Fleming, "Overview of automotive sensors," IEEE Sensors Journal, vol. 1, no. 4, pp. 296–308, 2001.
7. N. Menachemi and T. H. Collum, "Benefits and drawbacks of electronic health record systems," Risk Management and Healthcare Policy, vol. 11, pp. 47–55, 2011.
8. I. Yaqoob, K. Salah, R. Jayaraman and Y. Al-Hammadi, "Blockchain for healthcare data management: Opportunities, challenges, and future recommendations," Neural Computing and Applications, vol. 7, pp. 1–6, 2021.
9. A. Giakoumaki, S. Pavlopoulos and D. Koutsouris, "Secure and efficient health data management through multiple watermarking on medical images," Medical and Biological Engineering and Computing, vol. 44, pp. 619–631, 2006.
10. P. T. Chen, C. L. Lin and W. N. Wu, "Big data management in healthcare: Adoption challenges and implications," International Journal of Information Management, vol. 53, p. 102078, 2020.
11. T. M. Fernandez-Carames and P. Fraga-Lamas, "Towards post-quantum blockchain: A review on blockchain cryptography resistant to quantum computing attacks," IEEE Access, vol. 8, pp. 21091–21116, 2020.
12. N. R. Mosteanu and A. Faccia, "Fintech frontiers in quantum computing, fractals, and blockchain distributed ledger: Paradigm shifts and open innovation," Journal of Open Innovation: Technology, Market, and Complexity, vol. 7, no. 1, p. 19, 2021.
13. Y. Zhao, Y. Ye, H. L. Huang, Y. Zhang, D. Wu, H. Guan, Q. Zhu, Z. Wei, T. He, S. Cao and F. Chen, "Realization of an error-correcting surface code with superconducting qubits," Physical Review Letters, vol. 129, no. 3, p. 030501, 2022.
14. J. R. Friedman, V. Patel, W. Chen, S. K. Tolpygo and J. E. Lukens, "Quantum superposition of distinct macroscopic states," Nature, vol. 406, no. 6791, pp. 43–46, 2000.
15. N. Laflorencie, "Quantum entanglement in condensed matter systems," Physics Reports, vol. 646, pp. 1–59, 2016.

16. K. Heshami, D. G. England, P. C. Humphreys, P. J. Bustard, V. M. Acosta, J. Nunn and B. J. Sussman, "Quantum memories: Emerging applications and recent advances," Journal of Modern Optics, vol. 63, no. 20, pp. 2005–2028, 2016.
17. W. J. Munro, A. M. Stephens, S. J. Devitt, K. A. Harrison and K. Nemoto, "Quantum communication without the necessity of quantum memories," Nature Photonics, vol. 6, no. 11, pp. 777–781, 2012.
18. H. J. Briegel, W. Dür, J. I. Cirac and P. Zoller, "Quantum repeaters: The role of imperfect local operations in quantum communication," Physical Review Letters, vol. 81, no. 26, p. 5932, 1998.
19. J. L. Brylinski and R. Brylinski, "Universal quantum gates," Mathematics of Quantum Computation, 2002, vol. 79.
20. D. P. DiVincenzo, "Quantum gates and circuits," Proceedings of the Royal Society of London, Series A: Mathematical, Physical and Engineering Sciences, vol. 454, no. 1969, pp. 261–276, 1998.
21. G. Burkard, D. Loss and D. P. DiVincenzo, "Coupled quantum dots as quantum gates," Physical Review B, vol. 59, no. 3, p. 2070, 1999.
22. D. Jaksch, J. I. Cirac, P. Zoller, S. L. Rolston, R. Côté and M. D. Lukin, "Fast quantum gates for neutral atoms," Physical Review Letters, vol. 85, no. 10, p. 2208, 2000.
23. S. Khezr, M. Moniruzzaman, A. Yassine and R. Benlamri, "Blockchain technology in healthcare: A comprehensive review and directions for future research," Applied Sciences, vol. 9, p. 1736, 2019.
24. I. Abu-Elezz, A. Hassan, A. Nazeemudeen, M. Househ and A. Abd-Alrazaq, "The benefits and threats of blockchain technology in healthcare: A scoping review," International Journal of Medical Informatics, vol. 142, p. 104246, 2020.
25. W. J. Gordon and C. Catalini, "Blockchain technology for healthcare: Facilitating the transition to patient-driven interoperability," Computational and Structural Biotechnology Journal, vol. 16, pp. 224–230, 2018.
26. S. Bhattacharya, A. Singh and M. M. Hossain, "Strengthening public health surveillance through blockchain technology," AIMS Public Health, vol. 6, no. 3, p. 326, 2019.
27. P. Dutta, T. M. Choi, S. Somani and R. Butala, "Blockchain technology in supply chain operations: Applications, challenges and research opportunities," Transportation Research Part E: Logistics and Transportation Review, vol. 142, p. 102067, 2020.
28. K. N. Griggs, O. Ossipova, C. P. Kohlios, A. N. Baccarini, E. A. Howson and T. Hayajneh, "Healthcare blockchain system using smart contracts for secure automated remote patient monitoring," Journal of Medical Systems, vol. 42, pp. 1–7, 2018.
29. A. Sharma, Sarishma, R. Tomar, N. Chilamkurti and B. G. Kim, "Blockchain based smart contracts for internet of medical things in e-healthcare," Electronics, vol. 9, no. 10, p. 1609, 2020.
30. I. A. Omar, R. Jayaraman, M. S. Debe, K. Salah, I. Yaqoob and M. Omar, "Automating procurement contracts in the healthcare supply chain using blockchain smart contracts," IEEE Access, vol. 9, pp. 37397–37409, 2021.
31. B. Bhushan, P. Sinha, K. M. Sagayam and J. Andrew, "Untangling blockchain technology: A survey on state of the art, security threats, privacy services, applications and future research directions," Computers & Electrical Engineering, vol. 90, p. 106897, 2021.
32. R. Ur Rasool, H. F. Ahmad, W. Rafique, A. Qayyum, J. Qadir and Z. Anwar, "Quantum computing for healthcare: A review," Future Internet, vol. 15, no. 3, p. 94, 2023.
33. T. Liu, D. Lu, H. Zhang, M. Zheng, H. Yang, Y. Xu, C. Luo, W. Zhu, K. Yu and H. Jiang, "Applying high-performance computing in drug discovery and molecular simulation," National Science Review, vol. 3, no. 1, pp. 49–63, 2016.
34. A. M. Fathollahi-Fard, A. Ahmadi and B. Karimi, "Sustainable and robust home healthcare logistics: A response to the COVID-19 pandemic," Symmetry, vol. 14, no. 2, p. 193, 2022.

35. Y. Han, I. Ali, Z. Wang, J. Cai, S. Wu, J. Tang, L. Zhang, J. Ren, R. Xiao, Q. Lu and L. Hang, "Machine learning accelerates quantum mechanics predictions of molecular crystals," Physics Reports, vol. 934, pp. 1–71, 2021.
36. Z. Zhao, X. Li, B. Luan, W. Jiang, W. Gao and S. Neelakandan, "Secure Internet of Things (IoT) using a novel Brooks Iyengar quantum Byzantine Agreement-centered blockchain networking (BIQBA-BCN) model in smart healthcare," Information Sciences, vol. 629, pp. 440–455, 2023.
37. D. Lee and S. N. Yoon, "Application of artificial intelligence-based technologies in the healthcare industry: Opportunities and challenges," International Journal of Environmental Research and Public Health, vol. 18, no. 1, p. 271, 2021.

11 Quantum Blockchain-Based Healthcare

Merging Frontiers for Secure and Efficient Data Management

Kartick Sutradhar
Department of Computer Science and Engineering,
IIIT Sri City, Chittor, India

Ranjitha Venkatesh
Department of Computer Science and Engineering,
GITAM School of Technology Bangalore, India

Priyanka Venkatesh
Department of Computer Science and Engineering,
Presidency University, Bangalore, India

11.1 INTRODUCTION

The healthcare data revolution has propelled the medical industry into a transformative era, driven by the convergence of cutting-edge technologies and data-driven approaches. This revolution encompasses a profound shift in how healthcare data is collected, stored, analyzed, and utilized to improve patient care, enhance research endeavors, and streamline healthcare operations. The widespread adoption of electronic health records (EHRs), coupled with advancements in data analytics, artificial intelligence, and wearable devices, has enabled the generation of vast amounts of valuable health-related information [1]. This data deluge has empowered medical professionals with unprecedented insights into patient conditions, treatment outcomes, and disease trends, allowing for more precise diagnoses and personalized treatment plans. Moreover, the integration of genomics, telemedicine, and population health management has further amplified the potential of this data revolution, promising improved patient outcomes, accelerated drug discovery, and a deeper understanding of public health dynamics [2]. However, this revolution is not without its challenges, as concerns regarding data security, interoperability, and ethical considerations must be carefully navigated to fully harness the benefits of this data-driven paradigm shift in healthcare.

DOI: 10.1201/9781003449256-11

The healthcare data revolution holds immense promise, it also presents challenges, including the need to ensure data privacy, standardize data formats, integrate disparate systems, and address ethical considerations [3]. As technology continues to evolve, the healthcare industry will need to adapt and innovate to fully realize the potential benefits of the data revolution while navigating these challenges. Healthcare data management encounters a multitude of formidable challenges that arise from the intricate nature of medical information, stringent privacy concerns, technological limitations, and regulatory obligations. The safeguarding of patient privacy in the face of escalating cyber threats and unauthorized access remains a delicate balancing act. Furthermore, the lack of seamless interoperability among diverse healthcare systems impedes efficient data exchange and hinders comprehensive patient care [4]. Ensuring the accuracy and reliability of data poses a critical hurdle, as inaccuracies can lead to misdiagnoses and compromised patient safety. Integrating data from disparate sources, such as EHRs, medical imaging, and wearable devices, necessitates standardized formats and protocols to overcome complexity [5]. Regulatory compliance, notably navigating frameworks, adds complexity to data handling. The challenge of data silos within healthcare institutions underscores the need for integrated systems to facilitate collaboration. The absence of uniform data standards and medical coding inhibits effective exchange and analysis. Financial constraints hinder the adoption of advanced data management tools and technologies, potentially leaving smaller healthcare providers at a disadvantage [6]. Ethical considerations and the responsible use of data for research and machine learning applications demand thoughtful navigation. Bridging the gap between healthcare expertise and technological proficiency among professionals is yet another dimension of this intricate landscape. Addressing these challenges collectively is imperative to unlock the full potential of healthcare data management amidst the ongoing digital transformation.

The emergence of quantum computing and blockchain technology has ushered in a new era of innovation and disruption across various industries [7]. Quantum computing, with its ability to perform complex calculations at speeds unattainable by classical computers, holds the promise of revolutionizing fields such as cryptography, optimization, and drug discovery. Its potential to solve intricate problems could lead to breakthroughs in medical research, financial modeling, and supply chain optimization. However, the technology is still in its nascent stages, facing challenges such as error correction and scalability before its full potential can be realized. On the other hand, blockchain technology has garnered attention for its decentralized and tamper-resistant nature. Originally associated with cryptocurrencies like Bitcoin, blockchain's applications have expanded to diverse sectors, including healthcare, supply chain management, and digital identity verification. Its transparent and secure ledger system enhances data integrity and streamlines processes, reducing fraud and enabling more efficient transactions. Yet, blockchain also confronts obstacles such as scalability, energy consumption, and regulatory frameworks, which must be addressed for widespread adoption. As quantum computing and blockchain continue to evolve, their convergence could yield synergistic outcomes, particularly in enhancing the security of blockchain networks through quantum-resistant cryptography. This collaboration could lead to more robust and secure data management and transactions, with implications across finance, healthcare, and beyond.

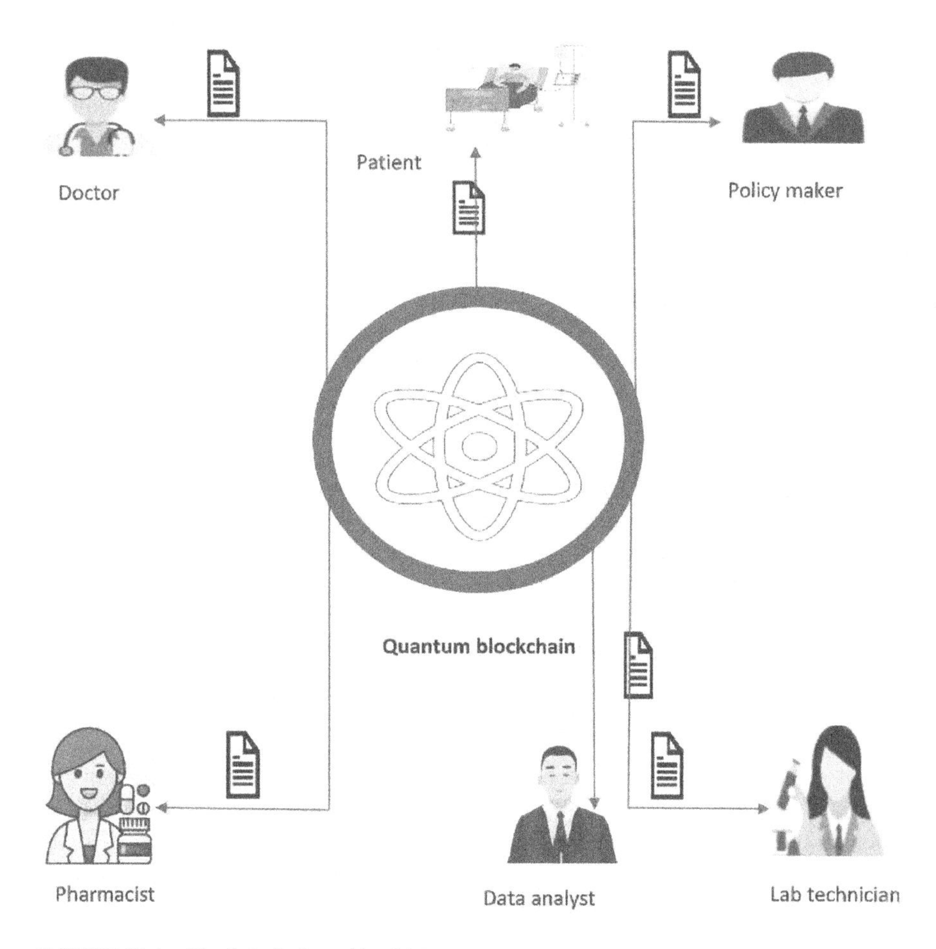

FIGURE 11.1 Blockchain-based healthcare system.

However, this convergence also brings forth new challenges and uncertainties, underscoring the need for ongoing research, collaboration, and thoughtful consideration of the potential risks and rewards. The quantum blockchain-based healthcare system can be shown in Figure 11.1.

11.2 FOUNDATIONAL CONCEPTS

11.2.1 Qubit Technology

Qubit technology represents a groundbreaking advancement in the field of quantum computing and quantum information processing [8]. Unlike classical bits, which can exist in a state of either 0 or 1, qubits have the unique property of existing in a superposition of both states simultaneously, due to the principles of quantum mechanics. This characteristic allows qubits to perform complex calculations and solve problems that would be practically impossible for classical computers. Additionally, qubits can be entangled, a phenomenon where the state of one qubit becomes dependent on the state of another, regardless of distance. This property holds promise for the development of highly secure quantum communication protocols. However, qubits are incredibly delicate and susceptible to environmental interference, making their stable manipulation a significant technical challenge. Researchers are exploring various physical implementations for qubits, such as superconducting circuits, trapped ions, and topological qubits, with the aim of harnessing their extraordinary potential to revolutionize computing, cryptography, and various other scientific and technological domains.

11.2.2 Superposition

Superposition is a fundamental concept in quantum mechanics that defies classical intuitions about how particles behave [9]. It refers to the ability of quantum particles, such as electrons or photons, to exist in a combination of multiple states simultaneously. Unlike classical objects, which can only be in one state at a time, quantum particles can be in a state of superposition where they exhibit a blend of different properties or states. For instance, an electron can be in a superposition of spin-up and spin-down states, or a photon can be in a superposition of different polarizations. This property underpins the power of quantum computing, as qubits can leverage superposition to perform many calculations in parallel. When qubits are entangled—another quantum phenomenon—their superposed states become correlated, allowing for intricate interactions that enable complex computations and advancements in fields like cryptography and materials science. Superposition challenges our classical understanding of reality but also offers the potential to unlock previously unimaginable technological capabilities [10].

11.2.3 Quantum Entanglement

Quantum entanglement is a mesmerizing phenomenon in the realm of quantum physics where two or more particles become intrinsically connected in such a way that the state of one particle instantaneously influences the state of another, regardless of the physical distance between them. This correlation is established when particles are created or interact in such a manner that their properties, like spin or polarization, become intertwined and interdependent [11]. This peculiar connection persists even when the particles are separated by vast distances, seemingly defying the constraints of classical physics and challenging our intuitive notions of causality and locality. Quantum entanglement has profound implications for the development of quantum technologies, such as quantum computing and quantum cryptography, where entangled particles can be harnessed to perform computations faster than classical counterparts and to create unbreakable codes for secure communication. Despite its incredible potential, entanglement remains an enigmatic aspect of quantum mechanics, prompting ongoing research and exploration into its underlying principles and applications.

11.2.4 Quantum Memories

Quantum memories represent a pivotal concept in the field of quantum information science, aimed at addressing the challenge of preserving and manipulating delicate quantum states over extended periods. These states can encode information in forms like qubits, which are susceptible to rapid degradation due to their sensitivity to environmental disturbances. Quantum memories function as devices capable of storing these fragile quantum states reliably, thus enabling the coherent transfer and retrieval of quantum information on demand. This technology holds immense promise for quantum communication networks and quantum computing systems. Quantum memories could facilitate the creation of efficient and secure communication channels by storing and distributing entangled quantum states, enabling long-distance quantum teleportation and encrypted communication [12]. Additionally, they could enhance quantum computers by mitigating errors and allowing for the synchronization of qubits during complex computations. Researchers are exploring a variety of physical implementations for quantum memories, including solid-state systems, trapped ions, and atomic ensembles, with the aim of harnessing their potential to revolutionize information processing and communication in the quantum realm.

11.2.5 Quantum Repeaters

Quantum repeaters represent a revolutionary advancement in the field of quantum communication, addressing the critical challenge of transmitting quantum information over long distances while preserving its delicate quantum properties. In classical communication, signal loss and degradation can be mitigated using traditional repeaters, but in the quantum realm, the no-cloning theorem prevents straightforward copying of quantum states [13]. Quantum repeaters overcome this limitation by employing entanglement swapping techniques, allowing for the creation of entangled pairs of qubits between distant nodes. These pairs, known as entanglement links, serve as the building blocks for long-range quantum communication. By repeating the entanglement swapping process along the communication channel, quantum repeaters extend the effective range of entanglement distribution, enabling secure and efficient quantum key distribution (QKD) and even the realization of long-distance quantum teleportation. This technology holds immense promise for creating secure quantum communication networks that could revolutionize fields like cryptography and information exchange. However, quantum repeaters face substantial technical challenges due to the fragile nature of entanglement and the need to maintain coherence over multiple iterations, making ongoing research crucial for their successful implementation.

11.2.6 Quantum Gates

Quantum gates are fundamental operators used in quantum computing to manipulate the state of qubits, the quantum equivalent of classical bits [14]. These gates perform operations on qubits, altering their quantum properties like superposition and entanglement. Quantum gates are represented as matrices, with each matrix describing how a gate transforms the state of a qubit or a pair of qubits [15]. These gates play a crucial role in constructing quantum algorithms by enabling the manipulation of qubit states to perform various computations and solve specific problems.

There are several types of quantum gates, each designed to manipulate qubits in different ways to perform specific operations in quantum computing. Here are some common types of quantum gates:

1. Pauli-*X* Gate: Also known as the "bit-flip" gate, it flips the state of a qubit from $|0\rangle$ to $|1\rangle$ and vice versa. Mathematically, it corresponds to a 180° rotation around the *X*-axis of the Bloch sphere.
2. Pauli-*Y* Gate: This gate is the "bit and phase-flip" gate, rotating the state of a qubit from $|0\rangle$ to $|1\rangle$ with a phase change. It combines a bit flip and a phase flip. It corresponds to a 180° rotation around the *Y*-axis of the Bloch sphere [16].
3. Pauli-*Z* Gate: The "phase-flip" gate, it introduces a phase change in the qubit's state without altering its probability amplitudes. It corresponds to a 180° rotation around the *Z*-axis of the Bloch sphere.
4. Hadamard Gate: This gate creates superpositions by transforming the $|0\rangle$ state into an equal superposition of $|0\rangle$ and $|1\rangle$, and the $|1\rangle$ state into an equal superposition of $|0\rangle$ and $|1\rangle$. It corresponds to a 90° rotation around the *X*-axis followed by a 180° rotation around the *Z*-axis.
5. CNOT Gate (Controlled-NOT): A two-qubit gate that performs a NOT operation on the target qubit (flipping its state) only if the control qubit is in state $|1\rangle$. It is a fundamental gate for creating entanglement and performing quantum error correction [17].
6. Toffoli Gate: Also known as the CCNOT gate (controlled–controlled-NOT), this gate acts on three qubits and performs a NOT operation on the target qubit if both control qubits are in state $|1\rangle$.
7. SWAP Gate: A gate that swaps the states of two qubits. It's useful for rearranging qubit states within a quantum circuit.
8. Phase Gate: This gate introduces a phase shift to the qubit state, allowing manipulation of the quantum phase without altering probability amplitudes.
9. Rotation Gates: These gates allow rotation of the qubit state around various axes of the Bloch sphere by specific angles, enabling precise control of quantum information.

These gates, when combined in specific sequences, form quantum circuits that perform quantum computations. Each gate contributes to the unique computational power of quantum computers, enabling them to solve problems that are challenging for classical computers.

11.3 BLOCKCHAIN TECHNOLOGY IN HEALTHCARE

Blockchain technology, at its core, is a decentralized and immutable digital ledger system that enables secure and transparent record-keeping of transactions. It operates through a network of computers, or nodes, where each transaction is added to a "block" of data, which is then linked to the previous block in chronological order, forming a "chain." This structure ensures that once data is recorded, it cannot be altered without consensus from the majority of the network, making blockchain highly resistant to tampering and fraud [18]. The technology's decentralized nature eliminates the need for intermediaries, reducing costs and enhancing efficiency in various processes, from financial transactions to supply chain management and beyond [19]. Smart contracts, self-executing programs with predefined rules, can also be deployed on blockchains, automating actions when specified conditions are met. While blockchain holds immense potential for data security, transparency, and decentralized applications, challenges such as scalability, energy consumption, and regulatory considerations remain as the technology continues to evolve and find broader applications across industries [20].

11.3.1 Decentralization and Immutability

Decentralization and immutability are two foundational principles that underpin the power and uniqueness of blockchain technology. Decentralization refers to the distribution of authority and control across a network of nodes, rather than being concentrated in a single central entity. This architecture ensures that no single participant has complete control over the entire system, reducing the risk of manipulation or unauthorized alterations. Immutability, on the other hand, refers to the unchangeable nature of data once it is recorded on a blockchain [21]. Each transaction is cryptographically linked to the previous one, forming an unbroken chain of blocks. Once data are added to a block and the block is added to the chain, it becomes extremely difficult to alter or delete that data without the consensus of the majority of the network. This immutability not only enhances data integrity and security but also fosters trust among participants, making blockchain particularly valuable in contexts where transparent and tamper-resistant record-keeping is essential, such as financial transactions, supply chain management, and digital identity verification. The combination of decentralization and immutability sets blockchain apart as a powerful tool for enabling secure and accountable interactions in a wide range of industries and applications [22].

11.3.2 Smart Contracts for Healthcare

Smart contracts hold great promise for revolutionizing healthcare by automating and enhancing various aspects of medical processes and data management. These self-executing digital contracts are built on blockchain technology and enable the automatic execution of predefined actions when specific conditions are met. In healthcare, smart contracts could streamline administrative tasks, such as insurance claims processing and billing, by automating verification and payment processes, reducing paperwork and processing times [23]. Additionally, these contracts could facilitate secure and transparent sharing of patient data between different healthcare providers, ensuring that only authorized parties have access and enhancing data privacy. Smart contracts could also improve clinical trials by automating consent processes, tracking trial milestones, and disbursing payments to participants based on predefined criteria. Moreover, in telemedicine and remote patient monitoring, smart contracts could automate appointment scheduling, monitor patient data, and trigger alerts or interventions when certain health metrics cross predefined thresholds [24]. Despite the potential benefits, implementing smart contracts in healthcare requires addressing challenges like legal and regulatory considerations, standardization of data formats, and ensuring the accuracy of input data. With proper safeguards and development, smart contracts have the potential to optimize healthcare operations, enhance patient care, and transform the way the industry manages administrative and clinical processes [25].

11.3.3 Privacy and Security Challenges in Traditional Blockchains

Privacy and security challenges have emerged as significant concerns within traditional blockchain systems. While the transparent and immutable nature of blockchain is advantageous in many contexts, it poses inherent risks to sensitive information in certain applications. Traditional blockchains, such as the Bitcoin and Ethereum networks, store all transaction data on a public ledger, which means that once data is recorded, it cannot be altered. However, this openness compromises privacy by allowing anyone to access and trace transaction details, potentially revealing sensitive financial, personal, or business information. Furthermore, the pseudonymous nature of blockchain addresses does not guarantee true anonymity, as sophisticated analysis techniques can potentially link addresses to real-world identities [26].

Additionally, the decentralized and distributed nature of traditional blockchains can result in slower transaction speeds and higher energy consumption due to the consensus mechanisms used to validate transactions. These challenges hinder the scalability and efficiency needed for widespread adoption in high-demand environments like healthcare, where real-time processing and privacy of patient data are crucial. Efforts are underway to address these issues. Some blockchain projects are exploring techniques like zero-knowledge proofs, ring signatures, and secure multi-party computation to enhance privacy while preserving the benefits of decentralization. As blockchain technology evolves, finding effective solutions to these privacy and security challenges will be essential to unlocking its potential across various industries while safeguarding sensitive data.

11.4 QUANTUM COMPUTING IN HEALTHCARE

Quantum computing holds the potential to revolutionize healthcare by tackling complex problems that are beyond the capabilities of classical computers. Its immense processing power could accelerate drug discovery processes by simulating molecular interactions and predicting the efficacy of potential compounds with unprecedented accuracy [27]. Quantum computing could optimize the development of personalized treatment plans by analyzing vast amounts of genomic and patient data, enabling more targeted therapies for individual patients. Moreover, quantum algorithms could enhance medical imaging techniques, such as MRI and CT scans, by processing data faster and producing higher resolution images, leading to more accurate diagnoses. In drug delivery, quantum computing could optimize the design of nanoparticles for targeted therapies, improving drug effectiveness while minimizing side effects. Despite these promising applications, practical implementation is a challenge due to the fragile nature of quantum states and the need for error correction. As quantum computing continues to advance, collaboration between quantum experts and healthcare professionals will be essential to fully harness its potential for transformative advancements in medical research, diagnosis, and treatment.

11.4.1 Drug Discovery and Molecular Simulations

Drug discovery and molecular simulations have been profoundly transformed by advanced computational techniques, ushering in a new era of efficiency and innovation in pharmaceutical research. Through sophisticated modeling and simulations, researchers can virtually explore the interactions between drug compounds and biological molecules at a molecular level, expediting the identification of promising drug candidates. This process significantly accelerates the drug discovery pipeline by reducing the need for extensive laboratory experiments, which can be time-consuming and expensive. Molecular simulations enable the analysis of complex molecular structures, predicting how potential drugs will bind to target proteins and understanding their effects on cellular processes. These simulations provide valuable insights into drug mechanisms, aiding in the design of safer and more effective pharmaceuticals [28].

11.4.2 Optimization in Healthcare Logistics

Optimization techniques are playing a pivotal role in revolutionizing healthcare logistics, enhancing the efficiency and effectiveness of various processes within the healthcare supply chain. From the procurement of medical supplies to the distribution of medications and the management of hospital resources, optimization strategies are enabling healthcare organizations to streamline operations, reduce costs, and improve patient care. By leveraging advanced algorithms and data-driven insights, healthcare logistics can better forecast demand, allocate resources, and optimize routing and scheduling [29].

One significant area where optimization is making a profound impact is inventory management. Healthcare facilities must maintain a delicate balance between having adequate stock to meet patient needs while minimizing excess inventory and associated costs. Optimization algorithms analyze historical consumption patterns, external factors, and lead times to determine optimal reorder points and quantities. This ensures that critical medical supplies, such as medications and equipment, are available when needed, reducing stockouts and waste. Furthermore, optimization techniques optimize the routing and scheduling of deliveries, particularly crucial for time-sensitive items like medications, blood products, and organs for transplantation. By considering factors like distance, traffic patterns, and delivery deadlines, healthcare organizations can ensure timely and efficient distribution, enhancing patient care and potentially saving lives.

Optimization is also contributing to better resource allocation within healthcare facilities. From managing hospital beds to scheduling operating room procedures, optimization algorithms optimize resource utilization, minimizing bottlenecks and improving patient flow. This leads to reduced wait times for patients, improved utilization of healthcare professionals' time, and ultimately better patient outcomes. However, challenges remain in implementing optimization strategies in healthcare logistics. Integration with existing information systems, data accuracy, and resistance to change are hurdles that need to be addressed. Moreover, ethical considerations, such as prioritizing certain patients or resources, require careful navigation. Despite these challenges, the integration of optimization techniques in healthcare logistics holds tremendous promise for improving resource management, patient care, and overall operational efficiency. As technology continues to advance and healthcare organizations increasingly embrace data-driven decision-making, optimization will remain a cornerstone of transforming healthcare logistics for the better.

11.4.3 Artificial Intelligence and Machine Learning Acceleration

AI and machine learning are experiencing a rapid acceleration that is reshaping industries and pushing the boundaries of innovation. With the exponential growth in computing power, availability of large datasets, and advancements in algorithms, AI and machine learning have gained the ability to process and interpret complex information at an unprecedented scale and speed. These technologies are revolutionizing fields such as healthcare, finance, manufacturing, and more. In healthcare, AI is being used to diagnose diseases, predict patient outcomes, and discover new drug candidates. In finance, machine learning algorithms are enhancing fraud detection, algorithmic trading, and risk assessment. Industries are leveraging AI-powered chatbots, recommendation systems, and predictive analytics to enhance customer experiences and optimize operations. The acceleration of AI and machine learning is unlocking new possibilities and driving transformative changes that will continue to reshape how we live and work in the years to come. However, this acceleration also brings challenges such as ethical considerations, data privacy, and the need for responsible AI deployment, emphasizing the importance of a thoughtful and balanced approach to harnessing the full potential of these technologies [30].

11.5 CONVERGENCE OF QUANTUM AND BLOCKCHAIN IN HEALTHCARE

The convergence of quantum computing and blockchain technology holds immense promise for revolutionizing healthcare in unprecedented ways. Quantum computing's exceptional processing power could optimize complex medical simulations, revolutionizing drug discovery and molecular research. It could rapidly analyze vast datasets, such as genomics information, facilitating personalized medicine tailored to individual patients' genetic profiles. The integration of quantum-resistant cryptography within blockchain systems could bolster data security and privacy, vital for safeguarding sensitive patient information.

11.5.1 Quantum-secured Blockchain

Quantum-secured blockchain represents a cutting-edge fusion of quantum computing and blockchain technology, aimed at addressing the impending threat that quantum computers pose to traditional cryptographic systems. As quantum computers advance, they have the potential to break conventional encryption methods, compromising the security of data stored and transmitted on existing blockchains. Quantum-secured blockchain solutions leverage quantum-resistant cryptographic algorithms to ensure data integrity and privacy even in the face of quantum attacks. By integrating these advanced encryption techniques into blockchain networks, the technology offers an unprecedented level of security, making it resistant to hacking attempts from both classical and quantum computers. Quantum-secured blockchain holds significant implications for industries requiring high levels of data protection, such as finance, healthcare, and government, where sensitive information and transactions demand uncompromised security. However, challenges remain, including the development of efficient quantum-resistant algorithms and the integration of these solutions into existing blockchain architectures. Despite these obstacles, quantum-secured blockchain stands as a pioneering approach to safeguarding the future of data security in an era of rapidly advancing quantum computing technology [31].

11.5.2 Quantum Cryptography for Secure Transactions

Quantum cryptography emerges as a groundbreaking solution to fortify secure transactions within the healthcare domain, where data confidentiality and integrity are paramount [32]. Traditional cryptographic methods could become vulnerable to quantum attacks as quantum computers advance, jeopardizing patient privacy and the confidentiality of medical records. Quantum cryptography harnesses the principles of quantum mechanics to enable the exchange of encryption keys with an unprecedented level of security. QKD ensures that any eavesdropping attempts would disrupt the quantum state of the transmitted keys, thereby alerting authorized parties to potential breaches. In healthcare, this technology could be harnessed to secure EHRs, telemedicine sessions, and medical IoT devices. Patients' sensitive information could be safeguarded from unauthorized access, and medical professionals could share critical data with the confidence that it remains unaltered and tamper-proof. Quantum cryptography not only offers robust data protection but also has the potential to transform healthcare operations by establishing trust in digital interactions, enhancing patient data privacy, and paving the way for secure and seamless healthcare transactions in an increasingly interconnected and data-driven ecosystem. However, challenges, including the need for practical quantum cryptographic systems and standardized implementation, must be addressed before realizing its full potential in securing healthcare transactions.

11.5.3 Quantum-resistant Blockchain Consensus Algorithms

Quantum-resistant blockchain consensus algorithms are emerging as a crucial safeguard in healthcare to ensure the longevity and security of blockchain networks against the threat of quantum computers. Traditional consensus algorithms, such as Proof of Work or Proof of Stake, could be compromised by quantum attacks, potentially leading to unauthorized access or tampering of sensitive patient data. Quantum-resistant consensus algorithms, built on cryptographic primitives designed to withstand quantum computing capabilities, provide robust defense mechanisms. These algorithms, like the Quantum Byzantine Fault Tolerance or the lattice-based algorithms, can secure data by utilizing encryption methods that remain unbreakable even in the presence of quantum adversaries. By integrating these quantum-resistant consensus mechanisms into healthcare blockchains, the industry can ensure the integrity of patient records, medical research, and other critical data. This proactive approach not only safeguards against future quantum threats but also establishes a foundation for secure and reliable healthcare data management, enabling trust and transparency in an increasingly digitized healthcare landscape. Despite the promising potential of quantum-resistant blockchain consensus algorithms, practical implementation challenges and standardization efforts need to be overcome for seamless integration into healthcare systems, reinforcing the critical importance of ongoing research and collaboration in this dynamic field [33].

11.5.4 Ensuring Privacy with Quantum Encryption

Ensuring privacy in healthcare data has never been more critical, and the advent of quantum encryption brings a transformative solution to the forefront. Quantum encryption leverages the fundamental properties of quantum mechanics to create unbreakable encryption keys, providing an unprecedented level of security for sensitive patient information. Traditional encryption methods could potentially be vulnerable to the computational power of quantum computers, risking the exposure of medical records and private health data. Quantum encryption, however, establishes an unbreakable shield by using the laws of physics to detect any unauthorized attempts to intercept or tamper with data during transmission. This technology enables healthcare organizations to transmit EHRs, telemedicine communications, and other critical information with the confidence that it remains completely confidential and immune to eavesdropping. By adopting quantum encryption, the healthcare industry takes a proactive stance in safeguarding patient privacy, maintaining data integrity, and complying with stringent data protection regulations. As quantum technologies continue to evolve, integrating quantum encryption into healthcare systems has the potential to redefine data security standards, fortifying patient trust and advancing the seamless and secure exchange of medical information in an increasingly interconnected digital healthcare landscape [34].

11.6 CURRENT ADVANCES AND CASE STUDIES

There have not been any widely recognized or established instances of quantum blockchain being applied directly in the healthcare sector. While both quantum computing and blockchain technology have made significant advancements, their convergence into quantum blockchain is still in its early stages. Quantum computing has shown promise in various fields, such as cryptography, optimization, and scientific simulations, while blockchain has been extensively explored for its applications in data security and transparency. While there is great potential for the application of quantum blockchain in healthcare, such as secure patient data management, drug discovery, and medical research, practical implementations and concrete advances in this specific area might not have reached a level of widespread adoption or notable case studies.

A quantum protocol for a secure Internet of Things was proposed by Zhao et al. This study ensures the fairness and integrity of the interchange of medical data [35]. Based on the mutual authentication of Blum BlumShub and Okamoto Uchiyana Cryptosystem (OUCS), this study provides an OUCS-based mutual authentication system. Blockchain-based quantum IoT framework for safe medical data was discussed by Qu et al. This research develops a novel distributed quantum electronic medical record system and suggests a new private quantum blockchain network based on security issues [36]. This quantum blockchain's data structure uses entangled states to link the blocks together. Less storage space is needed because the time stamp is automatically generated by combining quantum blocks using specified activities. Each block's hash value is kept in a single qubit. The processing of quantum information is covered in great detail in the quantum electronic medical record protocol. A quantum blockchain-based system for a secure Internet of Things was introduced by Meng et al. A cutting-edge quantum blockchain-based medical data processing system (QB-IMD) is created in this work [37]. A quantum blockchain structure and a ground-breaking electronic medical record algorithm are introduced in QB-IMD to ensure the accuracy and impermeability of the processed data. A procedure for securing the medical image that can enhance the diagnosis process was proposed by Janani et al. It has been found that the suggested quantum block-based scrambling can increase the privacy-preserving process for medical photos [38]. In order to ensure the integrity of medical images, it also includes specific quantum encryption for ROI-based regional data. A strategy to enhance the diagnosis procedure by enhancing the medical image was introduced by Camphausen et al. The final product represents a crucial first step toward scaling real-world quantum imaging advantage and has the potential to be applied to basic research as well as biomedical and commercial applications [39]. A study on the use of quantum algorithms for drug discovery was discussed by Blunt et al. This paper presents novel estimates of the quantum computing cost of simulating progressively larger embedding sections of an important covalent protein–drug combination involving the drug Ibrutinib [40]. Additionally, they compare and quickly summarize the scaling characteristics of cutting-edge quantum algorithms. A quantum method for drug discovery was proposed by Mustafa et al. In order to study how this issue might be resolved using quantum computing and Qiskit Nature, this research compares the Variational Quantum Eigensolver and the Quantum Approximate Optimization Algorithm [41]. The comparison of existing protocols can be shown in Table 11.1.

11.7 BENEFITS AND CHALLENGES

Quantum blockchain technology presents a paradigm shift in healthcare, offering a plethora of advantages that have the potential to revolutionize data management, security, and collaboration. One key advantage lies in the unprecedented level of

TABLE 11.1
Comparison of Existing Protocols

Protocols	Quantum blockchain	Quantum attacks	Qubit technology	Quantum gates	Quantum memories	Quantum repeaters
Zhao et al. [35]	✓	✓	✓	✓	X	X
Qu et al. [36]	X	✓	✓	✓	X	✓
Meng et al. [37]	✓	✓	✓	✓	✓	X
Janani et al. [38]	✓	X	✓	✓	X	✓
Camphausen et al. [39]	✓	✓	✓	X	✓	X
Blunt et al. [40]	✓	✓	✓	✓	X	✓
Mustafa et al. [41]	✓	X	✓	✓	✓	X

data security provided by quantum encryption, safeguarding sensitive patient information against both classical and quantum computing threats. The fusion of quantum computing with blockchain enhances data processing capabilities, enabling complex medical simulations, accelerating drug discovery, and facilitating personalized treatment plans through advanced analytics on vast datasets. Quantum-resistant consensus algorithms fortify the integrity of healthcare transactions and records, ensuring immutability even in the face of quantum attacks. Moreover, the decentralized nature of blockchain combined with quantum principles enables secure and efficient sharing of patient data among healthcare providers, enhancing care coordination while preserving patient privacy [6]. By harnessing the power of quantum blockchain, healthcare systems can usher in a new era of precision medicine, secure data exchange, and streamlined operations, ultimately leading to improved patient outcomes and a more resilient healthcare infrastructure. However, while the advantages are substantial, the integration of quantum blockchain in healthcare requires careful consideration of technical, ethical, and regulatory challenges to fully realize its transformative potential [8].

While the fusion of quantum computing and blockchain technology holds tremendous promise for healthcare, it also presents notable challenges and limitations that must be carefully navigated. One significant challenge is the nascent stage of both quantum computing and quantum-resistant cryptography. Developing practical quantum computers and reliable quantum-resistant algorithms is an ongoing endeavor, and their integration into healthcare systems demands rigorous research and testing. Additionally, the highly specialized nature of quantum technology requires expertise that may not be readily available within healthcare organizations, potentially creating a shortage of skilled professionals to implement and maintain these systems effectively. Furthermore, the integration of quantum elements into blockchain networks introduces complexities. Quantum-resistant consensus algorithms must be developed to withstand quantum attacks while maintaining the decentralized and transparent nature of blockchain [9]. Ensuring compatibility and interoperability between quantum and existing blockchain infrastructure poses an additional hurdle. Data privacy is a paramount concern in healthcare, and while quantum encryption offers enhanced security, challenges remain. Key distribution and management in quantum encryption systems require careful consideration to prevent unauthorized access and maintain data confidentiality. Moreover, the potential threat of quantum attacks on classical blockchain encryption methods underscores the urgency to transition to quantum-resistant cryptographic mechanisms. Regulatory and ethical considerations also play a significant role. The adoption of quantum blockchain in healthcare will require adherence to evolving data protection regulations and ethical standards to ensure patient consent, data ownership, and secure handling of sensitive health information. The challenges and limitations of quantum blockchain in healthcare underscore the need for a comprehensive and multidisciplinary approach. While the potential benefits are substantial, addressing technical, operational, ethical, and regulatory hurdles is essential to harness the transformative potential of this convergence while upholding patient privacy and data security [10].

11.8 FUTURE DIRECTIONS OF QUANTUM BLOCKCHAIN IN HEALTHCARE

The future directions of quantum blockchain in healthcare are poised to bring about transformative changes in how medical data are managed, shared, and utilized. As quantum computing and quantum-resistant cryptography continue to mature, healthcare is likely to witness the widespread adoption of quantum blockchain systems that offer an unparalleled level of data security and privacy. These systems could revolutionize EHRs by enabling patients to have complete control over their health data through secure and transparent ownership facilitated by blockchain, while the quantum element ensures unbreakable encryption. Moreover, quantum blockchain could catalyze breakthroughs in medical research by processing vast amounts of genomic and clinical data at unprecedented speeds. This could lead to personalized treatment plans that are finely tailored to each patient's unique genetic makeup, improving treatment efficacy and patient outcomes. Supply chain management within healthcare could also benefit from quantum blockchain, ensuring the authenticity and quality of pharmaceuticals and medical equipment by providing an immutable record of their origins and handling [42–45].

Telemedicine is another area that could undergo significant transformation. Quantum blockchain could facilitate secure and private virtual consultations, enabling real-time interactions between patients and healthcare providers across geographical boundaries while ensuring data confidentiality. However, achieving these future directions requires overcoming substantial challenges, including technological refinement, scalability of quantum computing, and addressing ethical and regulatory considerations. Collaborative efforts among quantum scientists, blockchain developers, healthcare professionals, and policymakers will be essential to navigate these complexities and harness the full potential of quantum blockchain in healthcare. As these challenges are addressed, the integration of quantum blockchain is likely to usher in a new era of secure, patient-centric, and data-driven healthcare, fundamentally altering the way healthcare is delivered, experienced, and advanced.

11.9 CONCLUSION

This chapter provides an in-depth exploration of the potential synergy between quantum computing and blockchain technology in the context of healthcare. It presents both theoretical concepts and practical applications, aiming to shed light on the possibilities and challenges of this innovative approach to data management. By merging quantum capabilities with blockchain's decentralized structure, the healthcare industry stands to gain new avenues for data security, privacy, and interoperability, paving the way for a more efficient and patient-centric healthcare ecosystem. This chapter also provides the future directions of quantum blockchain technology in healthcare domain.

REFERENCES

1. A. Hoerbst and E. Ammenwerth, "Electronic health records," Methods of Information in Medicine, vol. 49, no. 04, pp. 320–336, 2010.
2. N. Menachemi and T. H. Collum, "Benefits and drawbacks of electronic health record systems," Risk Management and Healthcare Policy, vol. 11, pp. 47–55, 2011.
3. I. Yaqoob, K. Salah, R. Jayaraman and Y. Al-Hammadi, "Blockchain for healthcare data management: Opportunities, challenges, and future recommendations," Neural Computing and Applications, vol. 7, pp. 1–6, 2021.
4. A. Giakoumaki, S. Pavlopoulos and D. Koutsouris, "Secure and efficient health data management through multiple watermarking on medical images," Medical and Biological Engineering and Computing, vol. 44, pp. 619–631, 2006.
5. P. T. Chen, C. L. Lin and W. N. Wu, "Big data management in healthcare: Adoption challenges and implications," International Journal of Information Management, vol. 53, p. 102078, 2020.
6. T. M. Fernandez-Carames and P. Fraga-Lamas, "Towards post-quantum blockchain: A review on blockchain cryptography resistant to quantum computing attacks," IEEE Access, vol. 8, 21091–21116, 2020.
7. N. R. Mosteanu and A. Faccia, "Fintech frontiers in quantum computing, fractals, and blockchain distributed ledger: Paradigm shifts and open innovation," Journal of Open Innovation: Technology, Market, and Complexity, vol. 7, no. 1, p. 19, 2021.
8. Y. Zhao, Y. Ye, H. L. Huang, Y. Zhang, D. Wu, H. Guan, Q. Zhu, Z. Wei, T. He, S. Cao and F. Chen, "Realization of an error-correcting surface code with superconducting qubits," Physical Review Letters, vol. 129, no. 3, p. 030501, 2022.
9. J. R. Friedman, V. Patel, W. Chen, S. K. Tolpygo and J. E. Lukens, "Quantum superposition of distinct macroscopic states," Nature, vol. 406, no. 6791, pp. 43–46, 2000.
10. N. Laflorencie, "Quantum entanglement in condensed matter systems," Physics Reports, vol. 646, pp. 1–59, 2016.
11. K. Heshami, D. G. England, P. C. Humphreys, P. J. Bustard, V. M. Acosta, J. Nunn and B. J. Sussman, "Quantum memories: Emerging applications and recent advances," Journal of Modern Optics, vol. 63, no. 20, pp. 2005–2028, 2016.
12. W. J. Munro, A. M. Stephens, S. J. Devitt, K. A. Harrison and K. Nemoto, "Quantum communication without the necessity of quantum memories," Nature Photonics, vol. 6, no. 11, pp. 777–781, 2012.
13. H. J. Briegel, W. Dür, J. I. Cirac and P. Zoller, "Quantum repeaters: The role of imperfect local operations in quantum communication," Physical Review Letters, vol. 81, no. 26, p. 5932, 1998.
14. J. L. Brylinski and R. Brylinski, "Universal quantum gates," Mathematics of Quantum Computation, 2002, vol. 79.

15. D. P. DiVincenzo, "Quantum gates and circuits," Proceedings of the Royal Society of London, Series A: Mathematical, Physical and Engineering Sciences, vol. 454, no. 1969, pp. 261–276, 1998.
16. G. Burkard, D. Loss and D. P. DiVincenzo, "Coupled quantum dots as quantum gates," Physical Review B, vol. 59, no. 3, p. 2070, 1999.
17. D. Jaksch, J. I. Cirac, P. Zoller, S. L. Rolston, R. Côté and M. D. Lukin, "Fast quantum gates for neutral atoms," Physical Review Letters, vol. 85, no. 10, p. 2208, 2000.
18. S. Khezr, M. Moniruzzaman, A. Yassine and R. Benlamri, "Blockchain technology in healthcare: A comprehensive review and directions for future research," Applied Sciences, vol. 9, p. 1736, 2019.
19. I. Abu-Elezz, A. Hassan, A. Nazeemudeen, M. Househ and A. Abd-Alrazaq, "The benefits and threats of blockchain technology in healthcare: A scoping review," International Journal of Medical Informatics, vol. 142, p. 104246, 2020.
20. W. J. Gordon and C. Catalini, "Blockchain technology for healthcare: Facilitating the transition to patient-driven interoperability," Computational and Structural Biotechnology Journal, vol. 16, pp. 224–230, 2018.
21. S. Bhattacharya, A. Singh and M. M. Hossain, "Strengthening public health surveillance through blockchain technology," AIMS Public Health, vol. 6, no. 3, p. 326, 2019.
22. P. Dutta, T. M. Choi, S. Somani and R. Butala, "Blockchain technology in supply chain operations: Applications, challenges and research opportunities," Transportation Research, Part E: Logistics and Transportation Review, vol. 142, p. 102067, 2020.
23. K. N. Griggs, O. Ossipova, C. P. Kohlios, A. N. Baccarini, E. A. Howson and T. Hayajneh, "Healthcare blockchain system using smart contracts for secure automated remote patient monitoring," Journal of Medical Systems, vol. 42, pp. 1–7, 2018.
24. A. Sharma, Sarishma, R. Tomar, N. Chilamkurti and B. G. Kim, "Blockchain-based smart contracts for internet of medical things in e-healthcare," Electronics, vol. 9, no. 10, p. 1609, 2020.
25. I. A. Omar, R. Jayaraman, M. S. Debe, K. Salah, I. Yaqoob and M. Omar, "Automating procurement contracts in the healthcare supply chain using blockchain smart contracts," IEEE Access, vol. 9, pp. 37397–37409, 2021.
26. B. Bhushan, P. Sinha, K. M. Sagayam and J. Andrew, "Untangling blockchain technology: A survey on state of the art, security threats, privacy services, applications and future research directions," Computers & Electrical Engineering, vol. 90, p. 106897, 2021.
27. R. Ur Rasool, H. F. Ahmad, W. Rafique, A. Qayyum, J. Qadir and Z. Anwar, "Quantum computing for healthcare: A review," Future Internet, vol. 15, no. 3, p. 94, 2023.
28. T. Liu, D. Lu, H. Zhang, M. Zheng, H. Yang, Y. Xu, C. Luo, W. Zhu, K. Yu and H. Jiang, "Applying high-performance computing in drug discovery and molecular simulation," National Science Review, vol. 3, no. 1, pp. 49–63, 2016.
29. A. M. Fathollahi-Fard, A. Ahmadi and B. Karimi, "Sustainable and robust home healthcare logistics: A response to the COVID-19 pandemic," Symmetry, vol. 14, no. 2, p. 193, 2022.
30. Y. Han, I. Ali, Z. Wang, J. Cai, S. Wu, J. Tang, L. Zhang, J. Ren, R. Xiao, Q. Lu and L. Hang, "Machine learning accelerates quantum mechanics predictions of molecular crystals," Physics Reports, vol. 934, pp. 1–71, 2021.
31. E. O. Kiktenko, N. O. Pozhar, M. N. Anufriev, A. S. Trushechkin, R. R. Yunusov, Y. V. Kurochkin, A. I. Lvovsky and A. K. Fedorov, "Quantum-secured blockchain," Quantum Science and Technology, vol. 3, no. 3, p. 035004, 2018.
32. A. Abuarqoub, S. Abuarqoub, A. Alzu'bi and A. Muthanna, "The impact of quantum computing on security in emerging technologies," in The 5th International Conference on Future Networks & Distributed Systems, 2021, pp. 171–176.
33. W. Wang, Y. Yu and L. Du, "Quantum blockchain based on asymmetric quantum encryption and a stake vote consensus algorithm," Scientific Reports, vol. 12, no. 1, p. 8606, 2022.

34. H. Zhu, C. Wang and X. Wang, "Quantum fully homomorphic encryption scheme for cloud privacy data based on quantum circuit," International Journal of Theoretical Physics, vol. 60, pp. 2961–2975, 2021.
35. Z. Zhao, X. Li, B. Luan, W. Jiang, W. Gao and S. Neelakandan, "Secure Internet of Things (IoT) using a novel Brooks Iyengar quantum Byzantine Agreement-centered blockchain networking (BIQBA-BCN) model in smart healthcare," Information Sciences, vol. 629, pp. 440–455, 2023.
36. Z. Qu, Z. Zhang and M. Zheng, "A quantum blockchain-enabled framework for secure private electronic medical records in internet of medical things," Information Sciences, vol. 612, pp. 942–958, 2022.
37. Z. Qu, Y. Meng, B. Liu, G. Muhammad and P. Tiwari, "QB-IMD: A secure medical data processing system with privacy protection based on quantum blockchain for IoMT," IEEE Internet of Things Journal, 2023.
38. T. Janani and M. Brindha, "A secure medical image transmission scheme aided by quantum representation," Journal of Information Security and Applications, vol. 59, p. 102832, 2021.
39. R. Camphausen, A. S. Perna, Á Cuevas, A. Demuth, J. A. Chillón, M. Gräfe, F. Steinlechner and V. Pruneri, "Fast quantum-enhanced imaging with visible-wavelength entangled photons," Optics Express, vol. 31, no. 4, pp. 6039–6050, 2023.
40. N. S. Blunt, J. Camps, O. Crawford, R. Izsák, S. Leontica, A. Mirani, A. E. Moylett, S. A. Scivier, C. Sunderhauf, P. Schopf and J. M. Taylor, "Perspective on the current state-of-the-art of quantum computing for drug discovery applications," Journal of Chemical Theory and Computation, vol. 18, no. 12, pp. 7001–7023, 2022.
41. H. Mustafa, S. N. Morapakula, P. Jain and S. Ganguly, "Variational quantum algorithms for chemical simulation and drug discovery," in 2022 International Conference on Trends in Quantum Computing and Emerging Business Technologies (TQCEBT), 2022, pp. 1–8.
42. S. S. Gill, "Quantum and blockchain-based serverless edge computing: A vision, model, new trends and future directions," Internet Technology Letters, 2021, p. e275.
43. A. Bandyopadhyay, A. Sarkar, S. Swain, D. Banik, A. E. Hassanien, S. Mallik, A. Li and H. Qin, "A game-theoretic approach for rendering immersive experiences in the Metaverse," Mathematics, vol. 11, no. 6, p. 1286, 2023.
44. A. Sihna, H. Raj, R. Das, A. Bandyopadhyay, S. Swain and S. Chakrborty, "Medical education system based on Metaverse platform: A game theoretic approach," in IEEE 4th International Conference on Intelligent Engineering and Management (ICIEM 2023), 2023, pp. 1–6.
45. P. Gupta, K. Bhadani, A. Bandyopadhyay, D. Banik and S. Swain, "Impact of Metaverse in the near 'future'," in IEEE 4th International Conference on Intelligent Engineering and Management (ICIEM 2023), 2023, pp. 1–6.

12 Unstructured Data Handling in Healthcare Sector

Megha Motta
Department of Electronics and Communication Engineering, Jaypee University of Engineering and Technology, Guna, India

Partha Sarathy Banerjee
Department of Computer Science and Engineering, Jaypee University of Engineering and Technology, Guna, India

Deepak Sharma
Department of Electronics and Communication Engineering, Jaypee University of Engineering and Technology, Guna, India

12.1 INTRODUCTION

The knowledge of artificial intelligence (AI) came into reality in 1956 and has contributed to remarkable results in most of the segments over the duration of the past 12 years. With the amplified growth of AI, it has made remarkable results in the healthcare domain. There are multiple applications of AI, and the involvement of AI in the medical sector proves as one of the most critical applications with tremendous potential. It has made it much easier to review a number of records in less time duration providing faster treatments to the patients with high accuracy. It has the ability to learn very fast, analyze the situation or data, and finally make conclusions. They are also enabled with a quality of self-correctness that makes them much more accurate than humans [1]. AI has multiple applications for processing unstructured data like videos, photos, or doctor's notes. It provides support for making certain clinical decisions and contributes to the designing of an intelligent interface that helps enhance the patient's engagement at the clinics and maintains the records of the patient's response to the treatment. It also provides assistance in designing the predictive model for any hospital for maintaining the proper footfall of the patient and allocation of resources to the patient and his/her kith and kin.

In the healthcare sector, there is an ample amount of information that needs to be maintained and processed properly in order to benefit mankind. This information includes prescriptions, pathology reports, radiologists' reports, and discharge summaries of the patient. This valuable information is kept in healthcare systems in an unstructured format that makes it difficult for machines to access and compute it properly. However, the Natural Language Processing (NLP) technique comes up

 DOI: 10.1201/9781003449256-12

with the solution as it is capable of structuralizing the whole information from the unstructured information, determining its linguistic models, and then interpreting it into a language that is easily understood by the healthcare machines. NLP consists of different machine-learning models that have made it possible to support different languages around the globe. The involvement of NLP techniques in the healthcare domain has not only increased the quality but has also minimized the cost. NLP benefits the patient by giving proper knowledge about the illness either with the help of NLP classifiers or by using the pattern recognition technique.

Among various techniques of AI image processing, NLP has made the biggest contribution in the field of healthcare. Information and communication technology (ICT) also provides guidance for the progression of treatment, medical prescription, diagnosis, information retrieval, documentation, and communication to healthcare professionals. It is been designed with many updated technologies to make it a big success in the healthcare domain. NLP comes up with numerous implementations that enhance the facilities of medical ICT and also adapt the electronic medical record (EMR) system. In recent times, there are many applications that have come into existence for extracting valuable knowledge from the EMRs [2, 3]. If the information stored in EMRs is in the form of structured information, it will be processed with more ease and provide great support in the diagnosis of the patients, with the reduction of workload for the physicians. Nowadays, with the help of the sentiment analysis application of NLP, the information available on any social media platform, for example, Twitter, Facebook, and so on also assists in analyzing any disease dynamics and is gaining the most attention as per the research in recent studies [4, 5].

EMRs are manipulated and maintained with the help of image processing techniques. Any progression in the complex pathology situation is monitored with the help of medical images that provide assistance to physicians in treatment [6]. For extracting valuable information from any image of a medical prescription or doctor's note, image processing and NLP work as a combination and help healthcare professionals for diagnosing the illness [7]. For any particular prescription, it helps in evaluating the semantics of the information and measuring its accuracy as per the protocol of "Medical Request Service." It becomes very difficult for healthcare systems to realize the unstructured clinical data that is written in the form of narrative texts in EMRs. It is also required to monitor the various grammatical structures as there is a wide range of expressions articulated in miscellaneous natural languages. Therefore, it gives rise to a challenging situation for the timely availability of information needed for diagnosis and treatment. NLP also contributes to the process of decision-making by providing an aid to extract significant information from the text and convert it into structured data. NLP benefits the medical domain by reducing the cost of healthcare and also plays a vital role in the improvement of healthcare processes.

12.2 BACKGROUND

As the growth rate of the inhabitants is growing rapidly, there is a massive requirement for better healthcare facilities and healthcare professionals. It becomes difficult to handle a large number of patients without any assistance. Here comes the role of technical assistance that not only helps in processing the patient's information but also provides better treatment and even sometimes helps in diagnosing the illness of the patient. The evolution of AI has made a significant contribution to the healthcare sector. Among various techniques of AI, NLP has proved to be most effective for handling unstructured information and analyzing clinical data efficiently [8]. NLP came into the era of the 1950s, and then in the 1960s and 1970s, some very effective natural language systems like Eliza which was developed by Weizenbaum in the year 1966, SHRDLU designed by Winograd in 1972, and LUNAR by Woods in 1973 were introduced. These all were task-oriented with limited domain access applications [9, 10].

Dr. Naomii Sager a renowned pioneer of NLP introduced the project known as the Linguistic String Project and designed a parser of English and computer grammar that contributed to analyzing clinical reports as well [11]. Spyns later discussed some more NLP-based clinical systems that were designed to contribute further [12]. There are abundant publications that demonstrate different clinical NLP systems like SPRUS (that was further upgraded into Symtext and later on in MPLUS) [13]. Many medical systems like MedLEE and MetaMap were designed with the aim of extracting important information, structuring, and encoding the clinical details of patient reports, and converting them into automated processes [14, 15]. Carol Friedman designed MedLEE in association with the Biomedical Informatics and Radiology Departments at Columbia University. Another program known as MetaMap was designed at the National Library of Medicine by Dr. Alan Aronson for mapping the biomedical text to UMLS Metathesaurus. Certain clinical literature was also introduced that stated the symptoms of recognizing different disorder and their treatments [16]. Later on, research was also conducted on the semantics in addition to the syntactic level of linguistic analysis in order to process any natural language information with more accuracy. However, an ample amount of work is been demonstrated so far but this chapter will mostly focus on different applications of NLP in the healthcare domain for maintaining clinical records.

12.3 BASIC NLP MODULE FOR THE HEALTHCARE DOMAIN

NLP is best used to process unstructured data from different resources. In the healthcare domain, EMRs and social media are the major sources of unstructured information that help the healthcare professional to interpret the illness. As soon as the NLP converts unstructured information (text) into structured information, it becomes easy for healthcare systems to use it for categorizing patients, extracting insights, and finally summarizing the information. Figure 12.1 explains the basic module of the system. Standard EMRs that assemble the patient's information makes it slightly difficult for the healthcare systems to encounter some specific details about the patient. NLP establishes the interface with EMRs that makes it easier for healthcare professionals to encounter the information. In addition, it includes some associated words at the time of encounter with the concerned patients individually and also establishes the interface into segments. This makes it easier for healthcare professionals to recover the stored data easily and diagnose the patient with better treatment.

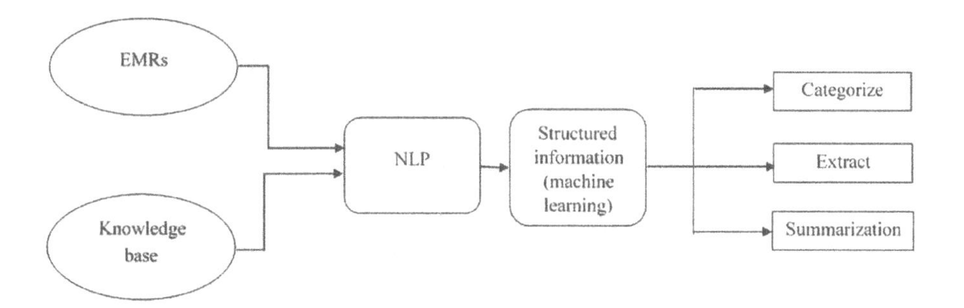

FIGURE 12.1 Fundamental NLP system for the healthcare domain.

NLP not only plays a vital role in maintaining EMRs but also enables predictive analysis that enhances the health concerns of the population. In recent research conducted in the United States, the problem of suicide attempts has increased with the passing of time. In 2018, an approach was designed that helped healthcare professionals to monitor social media accounts such as Twitter, Instagram, or Facebook to prevent suicide attempts with the help of sentiment analysis. Users who are prone to attempt suicide start posting fewer emojis in texts, emojis with a certain type of sadness or broken heart, and even sometimes start posting angry or frustrated texts before attempting suicide. This ability of NLP has made it more popular among recent technologies and has gained the maximum interest from researchers.

12.4 PROSPECTIVE APPLICATIONS OF NLP IN THE HEALTHCARE DOMAIN

In today's scenario where technology is also revising itself for better results, healthcare systems are also required to get upgraded to serve the best facility to mankind. Updated healthcare systems help clinicians by increasing their productivity and upskilling them. This advanced technology has also introduced some such applications that have served as a blessing. It helps in converting complicated medical terminologies into simpler words that can be easily understood by the patient. Below are some of the techniques that demonstrate the impact of NLP interventions in the healthcare domain.

12.4.1 Machine Translation Technology

It helps to convert text from one language to another language, for example, from English to Spanish. It is a Google Translate that is conceptualized on the concept of Statistical Machine Translation (MT) and uses multilingual statistical environments in combination with neural MT services. It has the ability to translate any text as well as website from one language to another. Its main function is to develop an algorithm that automatically translates information from one language to another without human intervention. In the medical sector, MT contributes to translating the complex medical phrases mentioned in the prescription into simple English language or the language that is easily understood by the patient's family members. This is possible without any human interference. MT effectively performs the task of translating medical reports and has set the healthcare domain to another level [17].

12.4.2 Speech Recognition Technology

It works on the concept of using voice as input and then performing the function of computing in the system. Nowadays, this technology has various applications in home automation, mobile phones, virtual assistance, video games, and so on. Its biggest and most vital application is voice assistants like Alexa, Siri, Google Assistant, and Cortana. Speech recognition software is designed with the help of multiple algorithms that comprise linear prediction (PLP) features, deep neural networks, Viterbi search, and so on. This technology uses voice recognition software using NLP algorithms. It is maximally used by healthcare professionals for dictating their prescriptions into their healthcare systems or even useful for updating the patient's EMRs. It benefits the whole healthcare domain by reducing the time for typing or selecting any particular text.

12.4.3 Sentiment Analysis or Opinion Mining

Sentiment analysis is also known as opinion mining or even emotional AI. It works with certain rules and falls under the category of data mining. It follows the set of instructions designed manually to perform the task of sentiment analysis. It works on the principle of analyzing the patient's feelings. It uses the NLP algorithm to extract information from any social media healthcare platform and then interprets the text data using analyzing techniques to specify the emotions whether it's positive, negative, or neutral. In the healthcare domain, it is best used for feedback services or reviews regarding healthcare services. It may also help new patients decide the best healthcare clinic for their illness depending upon the reviews they see on social media platforms with the help of sentiment analysis techniques. It not only contributes to the healthcare domain but also plays a vital role in business expansion. It contributes to the growth of any clinic on the basis of the patient's positive reviews and also informs the owner of the healthcare clinic to improve its services if there are any negative or neutral reviews.

12.4.4 Medical Question Answering (Q/A) System

In this technique of NLP, the user submits the question in the form of a natural language and gets an answer automatically by a Q/A system. It works with the aim of providing accurate answers to the questions. This technique works on the concept of maintaining a database which helps to generate answers quickly. There are several applications that work on this technique like Siri, virtual assistants, Alexa, different chat boxes, and Google Assistance. In the healthcare domain, it develops a system that automatically answers all the queries of the patient or any user asked in natural language. It also helps patients to inquire in depth about their illnesses and medical reports. It not only benefits the patient's point of view but also the healthcare professional in decision-making. As there is the massive availability of information in the form of research journals and on the web with different healthcare platforms, the use of Q/A systems has increased rapidly. Healthcare professionals as well as biomedical researchers frequently search regarding diseases, medicines, or any medical procedure to acquire accurate information. Therefore, the Q/A system delivers accurate answers to its users extracted from the source database rather than providing a list of source documents [18].

12.4.5 Medical Text Summarization

As the name suggests it takes one or more than one document as input and generates the output in the form of a single articulated text that summarizes the key points of all the input documents. It helps to analyze an ample amount of data from the source input by identifying the salient features of the text and presenting them automatically. It can be further classified as generic, that is, to compute any general information and query specific, that is, considering any particular information that needs to be summarized from the input documents. In the case of medical summarization, it summarizes and points out the medical information that helps to understand the strengths or weaknesses of any case. Therefore, the function of text summarization can be best explained as it helps to generate a summary of short and accurate answers from the large text document by focusing only on the key points of that document. It also reduces the reading time of healthcare professionals. To conduct the process of summarization smoothly, there are certain steps that are involved in this process. The first step is content selection, that is, to select the key points of any given input document. The second step comprises content organization, that is, identifying contradictions as well as removing the redundancy of text if found among the selected information. Last but not least step is known as content regeneration, that is, to produce the final output from the selected information in the form of natural language [19].

12.4.6 Automatic Clinical Text Segregation or Classification

Automatic clinical text classification helps to understand the information that is embedded in medical prescriptions. This technique works with the help of machine learning algorithms. Automated clinical text is designed to perform the task in two categories. The first one is a text classifier that uses symbolic techniques or statistical machine-learning techniques to classify the text. The second part is word embedding which takes place with the help of clinical abbreviation disambiguation, information retrieval, and named entity recognition (NER). Automatic clinical text classification functions by allocating some predefined categories with the help of the NER tagger to the whole text written in the document. Then, it automatically examines the content and assigns predefined categories to the content. This technique of NLP is most relevant for the patient. Nowadays, patients search for additional medical information related to their illness on the web. This condition takes place as it becomes difficult for a patient or their family members to understand the medical document they read. To get assured regarding their health condition, a large amount of the population in today's scenario search for their illness on the web. This sometimes helps in the favor of the patient as they can keep an eye on their own health situation and get better in less duration of time. Therefore, it performs the task of revealing the information from clinical text perceptions with the help of a mining technique. It helps the patient to understand the illness more accurately, for example, evaluation of the stage of cancer, specific characteristics of any disease, and pathological reports [20].

12.4.7 Optical Character Recognition

Optical character recognition is also known as optical character reader (OCR) or text recognition. This technique is used to convert any scanned document or image of a document into an editable and readable document with the help of a machine code. This application came into existence because the scanned documents can be easily read manually by any human but when it comes to machines the scanned document or text image is just like a series of black and white dots that is not recognized by the machine. OCR provides an aid for the machine to read the document and in turn, generate a soft copy of it. It works by analyzing the text of any scanned document and then decoding the characters into the form of a code that enables the machine to read it and convert it into any electronic format or generate a soft copy of it. Therefore, it can be easily explained by saying that OCR helps the computer to recognize all the words or characters of any scanned document or a digital image of any handwritten or typed document by means of the optical properties of that word or characters that are printed on the scanned document or image.

OCR software makes use of one of the two algorithms to recognize the character or word of any image or scanned document. The recognition of a word or character takes place one at a time. The NLP-based algorithms used are named pattern recognition and feature detection. It finds its main application in converting the information of clinical test reports or X-ray reports into the form of readable text formats.

12.4.8 Information Extraction

Information extraction (IE) is the application of NLP that is used maximally. It performs the function of automatically extracting the information in a structured format from any of the unstructured or semistructured documents or any other electronically available sources. Generally, they can also extract the information from the texts spoken in human language with the help of NLP. This activity is performed by analyzing the pattern in the text rather than performing the linguistic analysis. The extracted information is structuralized and can be used to perform multiple tasks in the healthcare domain. The use of IE can be very well seen in the biosurveillance sector. Here the information is in the form of symptoms that are extracted from the medical history or chief complaint field of a patient before admitting him to the emergency department of a hospital [21]. IE can also play a vital role in understanding the progression and prevalence of any epidemic by extracting the symptoms across many patients. IE has also laid its application in the pharmacovigilance sector by extracting the information in a structured format with the help of NLP from say *n* number of records of patients to analyze the reaction of any drug that they all are consuming. The process of IE can also take place with the help of NER or named entity normalization. In this technique, the information is categorized under certain labels and then the extraction process takes place. It is limited to certain labels such as the name of the patient, place, time, or any other type of label (maybe the name of medicine). This information is mapped with the standardized form of NER and the useful information is extracted easily. IE is also done by identifying the relationship between the two entities.

12.4.9 Information Retrieval (IR)

IR is the technique that is used to organize, store, retrieve, and evaluate the information derived from document repositories, especially in the form of text. The best example of IR can be seen when any user puts a query into the system. IR uses the concept of keywords to make its search more efficient. These keywords are the words that normally people use for searching on any search engine. This technique is mostly useful when the user has to access a large number of documents. In the healthcare domain, it helps the patient to retrieve information easily from EMRs or from the web. IR also uses NLP-based modified approaches that are also used in IE, that is, named entities for identifying complex information and evaluating the relationship between entities to obtain high accurateness while retrieving any details [22–24].

12.5 BASIC LIBRARIES USED BY NLP

The technology of NLP works with the help of certain libraries that provide support and necessary extensions to execute their function and enhance their performances. Some of the basic NLP libraries with their applications are demonstrated to help the healthcare domain improve its productivity.

12.5.1 Sentiment Analysis

Flair is the most powerful library that is available for sentiment analysis. It works on the deep learning approach for training the models and then predicting the sentiment for each word or token present in a sentence or within a document. Natural Language Tool Kit (NLTK) and TextBlob also have the capability to perform the analysis but they are rule-based classifiers.

12.5.2 Text Classification

Tensorflow Hub provides one of the best pretrained models of Embeddings from Language Models (ELMo)'s word embedding designed at Allen NLP. They are used to represent the background features of the input text. Learning of the model is done with the help of bidirectional Long Short Term Memory (LSTM). It performs the function of word embedding from GloVe designed by Stanford, Word2Vec introduced by Google, and FastText announced by Facebook.

12.5.3 Topic Modeling

This task is performed with the help of two libraries named Scikit Learn NMF also written as sklearn's NMF (nonnegative matrix factorization) and Gensim's LDA (Latent Dirichlet allocation). These are the libraries that are used to realize the topics that are present in the input data (corpus). Gensim is mostly used where there is a large corpus or can say ample amount of data to process. However, they have a limitation in that it does not include NMF which is also considered to find the topics in the text. Scikit Learn includes NMF as well as LDA to perform the task of topic modeling.

12.5.4 For Performing the Basic Applications of NLP

The NLTK is the main platform for designing Python projects that work on natural language spoken by humans. It has the ability to interface with more than 50 corpora (sets of documents) and several lexical assets. It has text-processing libraries to perform the functions of entity extraction, part-of-speech tagging, tokenization, parsing, semantic reasoning, and stemming.

12.6 FILE FORMATS USED TO STORE EMRs

The details of all the patients are stored in the form of a text document. If there are a large number of details or say the input consists of a large number of sentences then it is known as a Corpus. There are different file formats that are used by EMRs to store the details of the patient. These file formats are then processed and analyzed by NLP algorithms to produce the desired output. Sometimes the EMRs are stored in the Rich Text Format, if we consider the input corpus from the web then PubMed citations can be downloaded in Extensible Markup Language (XML) format, and full articles are available in Hypertext Markup Language file format. For sentiment analysis if the corpus is considered from Twitter or any social media, then it is in the XML-based Really Simple Syndication format. File formats used in NLP cannot be converted from one format to another for processing. There are several packages that are inherent in programming languages to process the different file formats.

12.7 SOME OF THE ALGORITHMS USED FOR COMPUTING DIFFERENT TASKS IN NLP

The implementation of the NLP technique depends upon the algorithms they are using to solve the specific problem in the healthcare domain. Algorithms are nothing but step-by-step instructions that are given to any machine to compute any specific task. Some of the significant NLP algorithms with their functions are explained in Table 12.1.

TABLE 12.1
Detailed Explanation of Different Tasks Performed in NLP with Their Specific Algorithms

S. No.	Algorithm name	Specific tasks
1.	CRF++ and Hidden Markov Model (HMM)	The operations performed with the help of these algorithms are chunking, part-of-speech (POS) tagging, and named entity extraction
2.	Maxent	This algorithm is used for the word alignment task in machine translation
3.	Edit Distance and Soundex	Both of them perform the function of a spelling checker
4.	Cocke Younger Kasami (CYK) and Chart Parsing	As the name suggests these algorithms perform the function of parsing in NLP
5.	Support vector machines (SVM) and naive Bayes	Both of them are applicable where there is the task of document classification
6.	Latent Dirichlet Allocation (LDA) and latent semantic indexing (LSI)	These algorithms are used for performing the operations of topic modeling and keyword extraction in NLP

12.8 CONCLUSION

NLP is the branch of AI that has the ability to make machines communicate in natural language that is spoken by humans. One of the biggest achievements of NLP is in the field of healthcare domain. It helps healthcare professionals in decision-making and also helps to translate the EMRs into natural language. It has also designed question-answering modules that are capable of answering patients' queries regarding their illnesses and delivering more information about the disease with great accuracy. In this chapter, various NLP-based healthcare applications are explained that contribute to the medical domain to enhance the productivity of clinicians and provide proper information to patients regarding their health. Most of the information related to healthcare is available in the form of an unstructured format that is not easily understood by the machine. Therefore, accessing this information in emergency situations becomes a challenge. NLP helps in the healthcare domain to extract important information from unstructured EMRs, analyze the grammatical structure of the record, and translate those complicated medical terms into a simple language that is easily understandable by machines to make any decision. This chapter has also shed some light on the basic libraries used in NLP and demonstrated different file formats that are used to store the EMRs in healthcare systems. Algorithms play the most vital role in the execution of any technique. This chapter comprises the list of algorithms that are used to perform the specific tasks of NLP. Thus, the author has tried to cover all the important aspects of NLP that make healthcare systems user-friendly, facilitate easy access to EMRs, and provide a platform for patients and healthcare professionals to communicate with the machine.

REFERENCES

1. Abid Haleem, Raju Vaishya, Mohd Javaid, Ibrahim Haleem Khan, Artificial Intelligence (AI) applications in orthopaedics: An innovative technology to embrace, Journal of Clinical Orthopaedics and Trauma, Volume 11, Supplement 1, 2020, Pages S80-S81, ISSN 0976-5662, https://doi.org/10.1016/j.jcot.2019.06.012.
2. V. Carchiolo, A. Longheu, M. Malgeri and G. Mangioni, “Multisource agent-based healthcare data gathering,” in Proceedings of FedCSIS, Sep. 2015, pp. 1723–1729. https://doi.org/10.15439/2015F302
3. Y. Si and K. Roberts, “A frame-based NLP system for cancer-related information extraction,” AMIA Annual Symposium Proceedings, vol. 2018, pp. 1524–1533, 2018.
4. V. Carchiolo, A. Longheu and M. Malgeri, “Using Twitter data and sentiment analysis to study diseases dynamics,” in Proceedings of ITBAM 2015, vol. 9267. New York, NY: Springer-Verlag New York, Inc., 2015, pp. 16–24. http://dx.doi.org/10.1007/978-3-319-22741-2_2
5. S. Doan, E. W. Yang, S. Tilak and M. Torii, “Using Natural Language Processing to extract health-related causality from Twitter messages,” in IEEE ICHI-W, June 2018, pp. 84–85. https://doi.org/10.1109/ICHI-W.2018.00031
6. B. Shickel, P. J. Tighe, A. Bihorac and P. Rashidi, “Deep EHR: A survey of recent advances in deep learning techniques for electronic health record (EHR) analysis,” IEEE Journal of Biomedical and Health Informatics, vol. 22, no. 5, pp. 1589–1604, 2018. https://dx.doi.org/10.1109/JBHI.2017.2767063
7. G. Litjens, “A survey on deep learning in medical image analysis,” Medical Image Analysis, vol. 42, pp. 60–88, 2017. https://doi.org/10.1016/j.media.2017.07.005
8. Banerjee, P. S. & Banerjee, J. (2020). NLP for Clinical Data Analysis: Handling the Unstructured Clinical Information. In D. Sisodia, R. Pachori, & L. Garg (Eds.), *Handbook of Research on Advancements of Artificial Intelligence in Healthcare Engineering* (pp. 359–367). IGI Global. https://doi.org/10.4018/978-1-7998-2120-5.ch018.
9. J. Weizenbaum, “ELIZA—A computer program for the study of natural language communication between man and machine,” Communications of the ACM, vol. 26, pp. 23–28. https://doi.org/10.1145/357980.357991
10. W. A. Woods, “Semantics and quantification in natural language question answering,” Advanced Computing, vol. 17, pp. 1–87, 1978.
11. S. Naomi, Natural Language Information Formatting: The Automatic Conversion of Texts to a Structured Data Base, in Advances in Computers, edited by Marshall C. Yovits, Elsevier, vol. 17, 1978, pp. 89–162, ISSN 0065-2458, ISBN 9780120121175, https://doi.org/10.1016/S0065-2458(08)60391-5
12. P. Spyns, “Natural Language Processing in medicine: An overview,” Methods of Information in Medicine, vol. 35, no. 4–5, pp. 285–301, 1996. PMID: 9019092.
13. A. J. Christensen, S. L. Ehlers, J. S. Wiebe, P. J. Moran, K. Raichle, K. Ferneyhough and W. J. Lawton. “Patient personality and mortality: A 4-year prospective examination of chronic renal insufficiency,” Health Psychology, vol. 21, no. 4, pp. 315–320, 2002. PMID: 12090673. https://doi.org/10.1037/0278-6133.21.4.315
14. Friedman D, Sunder S. Human Subjects. In: *Experimental Methods: A Primer for Economists.* Cambridge: Cambridge University Press; 1994:38–60. https://doi.org/10.1017/CBO9781139174176
15. W. C. Hahn and R. A. Weinberg, “Rules for making human tumor cells,” New England Journal of Medicine, vol. 347, pp. 1593–1603, 2002.
16. Aronson, A R. “Effective mapping of biomedical text to the UMLS Metathesaurus: the MetaMap program.” *Proceedings. AMIA Symposium* (2001): 17–21.

17. K. Papineni, S. Roukos, T. Ward and W. J. Zhu, "BLEU: A method for automatic evaluation of machine translation," in Proceedings of the 40th Annual Meeting on Association for Computational Linguistics, Association for Computational Linguistics, 2002, pp. 311–318.
18. D. Demner-Fushman and J. Lin, "Answering clinical questions with knowledge-based and statistical techniques," Computational Linguistics, vol. 33, pp. 63–103, 2007.
19. E. O'Carroll Bantum, "Machine learning for identifying emotional expression in text: Improving the accuracy of established methods," Journal of Technology in Behavioral Science, vol. 2, pp. 21–27, 2017.
20. D. Demner-Fushman, N. Elhadad and C. Friedman, "Natural Language Processing for health-related texts," in Biomedical Informatics: Computer Applications in Health Care and Biomedicine. Cham: Springer International Publishing, 2021, pp. 241–272.
21. G. Hripcsak, et al., "Syndromic surveillance using ambulatory electronic health records," Journal of the American Medical Informatics Association, vol. 16, no. 3, pp. 354–361, 2009.
22. A. Bandyopadhyay, A. Sarkar, S. Swain, D. Banik, A. E. Hassanien, S. Mallik, A. Li and H. Qin, "A game-theoretic approach for rendering immersive experiences in the Metaverse," Mathematics, vol. 11, no. 6, p. 1286, 2023.
23. A. Sihna, H. Raj, R. Das, A. Bandyopadhyay, S. Swain and S. Chakrborty, "Medical education system based on Metaverse platform: A game theoretic approach," in IEEE 4th International Conference on Intelligent Engineering and Management (ICIEM 2023), 2023, pp. 1–6.
24. P. Gupta, K. Bhadani, A. Bandyopadhyay, D. Banik and S. Swain, "Impact of Metaverse in the near 'future'," in IEEE 4th International Conference on Intelligent Engineering and Management (ICIEM 2023), 2023, pp. 1–6.

Index

S

T

U

V

W

Z

Printed in the United States
by Baker & Taylor Publisher Services